The Chemistry and Deposition of Nitrogen Species in the Troposphere

The Chemistry and Deposition of Nitrogen Species in the Troposphere

Edited by

A. T. Cocks
Technology and Environmental Centre, National Power plc,
Leatherhead

ROYAL
SOCIETY OF
CHEMISTRY

The Proceedings of a One-day Symposium organised by the Environment
Group of the Royal Society of Chemistry on 19 February 1992 at the Scientific
Societies' Lecture Theatre, London

Special Publication No. 115

ISBN 0-85186-355-8

A catalogue record for this book is available from the British Library

Published by The Royal Society of Chemistry,
Thomas Graham House, Science Park, Cambridge
CB4 4WF

Printed in Great Britain by Hartnolls Ltd, Bodmin

Preface

Nitrogen compounds play a central role in atmospheric chemistry and are associated with several of today's major environmental issues.

Nitrogen oxide emissions from combustion sources are involved in the complex atmospheric processes leading to the formation of photochemical oxidants. Such pollutants at ground level are responsible for photochemical smogs, and at higher altitudes in the free troposphere contribute to "greenhouse" warming.

Nitrogen oxides, ammonia, mainly from agricultural emissions, and their atmospheric reaction products, account for the major fraction of nitrogen deposition to terrestrial ecosystems. Excess nitrogen deposition may have adverse environmental consequences arising from acidification or eutrophication.

Current environmental legislation, based on European Community directives on emissions from large combustion sources and from petrol-driven motor vehicles, will bring about a general reduction in nitrogen oxide emissions in the UK and Europe over the next few years. This legislation will be reviewed before the end of the decade, and any further abatement is likely to be more targeted, with more stringent controls being required where they will produce the greatest environmental benefit. For this targeted approach to be worthwhile, it is essential that a sound quantitative link between emissions and environmental effects is established.

This book presents the proceedings of a symposium, organised by the Royal Society of Chemistry Environment Group in February 1992, with the aim of summarising the current knowledge and uncertainties in the atmospheric science linking emissions to the key environmental effects. Reviews of photochemical oxidants in the emission "plumes" of large combustion sources, and in urban and rural areas are presented. The critical loads concept, a method of quantifying ecosystem sensitivity, is also reviewed, together with particulate chemistry and wet deposition, key steps in the deposition process, and their synthesis in a large-scale nitrogen deposition model.

Whilst significant advances have been made in defining and understanding the physico-chemical processes governing the atmospheric behaviour of nitrogen species and their ecological effects, much research is still needed to quantify these adequately enough to permit a meaningful estimation of the results of proposed emission control strategies at the required spatial and temporal resolution.

The Royal Society of Chemistry would like to thank the authors for their contributions and their cooperation in producing their manuscripts in a form suitable for publication, and their affiliate organisations for permission to publish the papers herein.

Alan T. Cocks   *RSC Environment Group*

Contents

Photochemical Oxidants from Large Combustion Sources

P. A. Clark

NATIONAL POWER PLC, NATIONAL POWER TECHNOLOGY AND ENVIRONMENT CENTRE, KELVIN AVENUE, LEATHERHEAD KT22 7SE, UK

1 INTRODUCTION

It is well known that large combustion sources are not a significant direct source of photochemical oxidants. The title of this paper may thus be regarded as slightly misleading without further definition of the terms used.

'Photochemical oxidants' (PCOs) will be taken to include O_3 (by far the most abundant), NO_2, H_2O_2 and free radical species such as OH and HO_2 which (on the whole) may be considered to be photochemically derived from O_3. Where environmental issues are concerned 'from' may be regarded as carrying with it the element of 'blame', in that effects are often considered as arising 'from' a source if the effect would not occur if the source were absent, even if there is no primary link between the source and effect (e.g. it arises from a secondary pollutant). Lastly, 'large combustion source' can generally be taken to mean a power plant or incinerator which emits through a stack tall enough to avoid major effects from local buildings.

Large combustion sources have received particular attention in many areas of air pollution research primarily because they are easily identifiable as major emission sources and thus both easier to study and potentially easier to control (at least from the point of view of the number of units that require measures). The physical behaviour of emission plumes arising from such plant has been extensively studied, so that we are now in a position, at least under most circumstances, to make reasonably accurate predictions of the contribution that a particular primary emission of a substance such as SO_2 will make to local air quality. Similar modelling approaches can be made for most primary emissions, often with only minor correction for the particular chemical or physical characteristics of the emission. It is perhaps worth noting at this point that much of the work on relatively near field behaviour to be discussed below is of American origin and is motivated not only by an interest in species as potential toxic substances (NO_2 and O_3) but from an interest in the impact of NO_2 on plume visibility which, on the whole has been a much more active issue in the USA.

When one considers longer ranges or travel times, chemical transformation and scavenging play a much greater role in determining the atmospheric lifetime and hence the impact to any particular location. Hence, the particular chemical or physical characteristics of the emission are important variables, while the chemical character of the ambient atmosphere in which the emission is transported has an increasingly important role. Nevertheless, few, if any, circumstances may be identified in which the direct impact of an emission actually decreases as the strength of the emission increases. When considering the direct impact of primary emissions, therefore, the application of scientific understanding serves 'only' to determine the size of the environmental benefit to be gained from a particular control measure, not the sign. (For example, few people could be found who would argue that reduction of the SO_2 emission from a source would not lead to **some** reduction in sulphur deposition to a particular area, even if, in practice, that reduction could not be detected).

The situation is not so clearcut, however, in the case of photochemical oxidants (PCOs). The vast majority of PCOs that can be attributed to emissions arise as secondary pollutants generated from atmospheric chemical processes between several precursor species, possibly arising from different sources. Thus, the mere definition of a 'source' of PCOs becomes problematic, and the whole concept of 'source-receptor matrices' or, more colloquially, 'blame tables', so beloved of political debates, becomes meaningless. One is forced instead to assess the impact of proposed control scenarios *in toto* before deciding on an optimum strategy. Furthermore, and rather more importantly, the possibility exists for a negative effect of some emissions on PCO levels, at least in some circumstances and/or at some locations. The rather novel feature arises, therefore (unfamiliar to the politician and general public, if not the experienced scientist) that an 'air pollutant' may actually have a beneficial effect. The difficult task then arises of deciding how to balance these 'benefits' against any possible deficits arising from the emission.

A large power station may be continuously pumping emissions into the atmosphere equivalent to those from a small city through a hole only a few tens of square metres in area, with initial concentrations four or five orders of magnitude greater than those typical of ambient air. Four orders of magnitude dilution may occur before the plume impinges on the ground. This vast dynamic range of concentration means that, in problems where atmospheric chemistry plays an important role, the large combustion source forms a separate and distinct regime of behaviour which needs to be considered in detail before decisions can be made about its overall impact.

Apart from their obvious geometrical characteristics, the main feature which distinguishes large combustion sources is that, from the point of view of PCO formation, they consist almost entirely of oxides of nitrogen, the only other species present of major photochemical relevance being oxygen and water vapour. In practice, the majority of the NOx (generally at least 95%) leaves the stack in the form of NO. The concentration of reactive hydrocarbons may be regarded as negligible in the initial plume. As a result the plume is initially considerably less photochemically active than surrounding air. In addition the NO-rich plume acts as a major sink for ozone, thereby removing a major source of free radicals. The plume may therefore be regarded as a 'chemical desert' in comparison with its more verdant surroundings.

At the other extreme, there is no doubt that, eventually, the NOx originating from large combustion plant which survives loss and transformation processes will become indistinguishable from NOx arising from other sources, as the spatial scale of mixing greatly exceeds the typical separation of the sources. The surviving NOx will therefore contribute to the large scale generation of photoxidants in the same way as any other source (although the precise contribution will depend upon the fraction surviving other processes). This global scale PCO generation is beyond the scope of the current paper, but its possibility does suggest that there exists a general transition, as one moves from very local impact to global impact, from PCO destruction to (possible) PCO generation. The transition regime is likely to be somewhere in the meso to synoptic or regional scale.

The change in behaviour as one moves through the different regimes is of obvious importance to the assessment of the overall impact of emissions (and possible controls). The subject matter of this paper is thus a discussion of observations and model treatments of the differing behaviours and our understanding of the factors which determine them. As well as discussing the PCOs of primary interest, attention will also be focused upon the impact on PCOs such as H_2O_2 whose major importance is their impact upon other species such as SO_2 rather than as 'pollutants' in their own right.

2 LOCAL IMPACT OF NOx EMISSIONS

Emissions

It is not within the scope of this paper to discuss in detail the mechanisms and magnitude of NOx formation inside combustion plant. In general, the mechanism is rather more complex than that for a species such as SO_2, as the emission arises in part from (organic) nitrogen existing in the original fuel and in part from nitrogen in the combustion air. Emissions are thus rather more difficult to predict than SO_2 emissions, and depend upon the particular plant and combustion conditions obtaining. In practice the bulk emissions from large, modern power plant arise from full load operation and so are perhaps more predictable than might otherwise be the case. It should nevertheless be borne in mind that, as is the case in all air pollution issues, no prediction of impact can be more accurate than our knowledge of emissions.

The emission from a boiler is generally proportional to load for a given generator, although the proportionality factor may depend upon not only the physical characteristics of the boiler but also the operating conditions obtaining at any particular time and the fuel nitrogen. A typical figure for a 500 MW coal fired boiler without low NOx burners is a concentration of 400-500 ppm NOx in the flue gas. Low NOx burners reduce this by typically 40%.

The vast majority of this NOx is in the form of NO. The NO_2 content is generally rather less than 5% and is consequently very difficult to measure accurately using conventional methods which rely on the reduction of NO_2 to NO and subsequent detection by using chemiluminescent reaction with O_3. The primary NO_2 thus forms a very small (but possibly significant in the near field) contribution to ambient NO_2. The bulk of NO_2 formed in the plume results from subsequent chemical reactions which are detailed below.

Chemistry

NOx chemistry is fundamental to atmospheric chemistry in general, in that NO_2 photolysis forms a vital source of O_3 in the troposphere. In addition to primary NO_2 arising from the boiler, in the initial stages of plume dispersion NO_2 is formed from NO by two routes. The first is reaction with O_3:

$$NO + O_3 \rightarrow NO_2 \qquad (1)$$

The rate constant for this is k_1 = 55.91 exp(-1430/T) ppm^{-1}s^{-1} (Cocks and Fletcher, 1988 and references therein). Direct reaction with oxygen proceeds by a termolecular reaction, viz

$$2NO + O_2 \rightarrow 2NO_2 \qquad\qquad (2)$$

The rate constant for this reaction is k_2 = 2.143x10^{-12} exp(530/T) ppm^{-2}s^{-1} (Cocks and Fletcher, 1988). This reaction is of very little importance in the ambient atmosphere as it is very rapidly quenched by dilution. Inside a stack, with ~5% O_2 and 500 ppm NO, NO_2 is produced at a rate of about 0.1 ppm s^{-1} (at stack temperatures), so that only a small contribution is made to 'primary' NO_2 by this reaction inside the stack. The precise contribution inside the plume depends very much upon the rate of dilution (Varey et al, 1978). However, it is clear that at concentrations representative of even the most polluted air (say 1 ppm NO) the rate of NO_2 production by this route is very slow (e.g. only of order 10 ppb h^{-1} even at 1 ppm NO).

It is possible to distinguish the contribution from the two NO_2 formation routes because in (1) for every molecule of NO_2 produced an O_3 molecule is destroyed. Janssen and Elshout (1987) generally show a greater amount of NO_2 formed than O_3 lost in Dutch plumes, by as much as 30%, but this appears to be an unusual observation. Other authors generally find less than 10% (e.g. White, 1977; Meagher et al, 1981; Melo and Stevens, 1981; Richards et al 1981). It is possible that the plumes studied in the Netherlands were undergoing rather slower dilution than those generally studied in the USA, although few authors actually quote plume dispersion data to enable this to be verified.

The half life for the reaction between O_3 and NO is shown in Fig. 1 as a function of O_3 and NO concentration. For those sections of the graph where initial NO exceeds initial O_3 (so that the final state would be zero O_3) the time for half the O_3 to be destroyed is plotted, while for the sections where the reverse is true the time for half the NO to be destroyed is plotted. It is clear that at NO concentrations typical of ambient air the reaction has a timescale of one to a few minutes, while in a recently emitted plume the reaction is even more rapid, having a timescale of only seconds at typical ambient O_3 concentrations. Any O_3 which mixes into the plume is thus expected to be rapidly destroyed as soon as it comes into molecular contact with NO. However, two mixing timescales must be considered: the rate of molecular diffusion and the rate of turbulent dispersion. In the initial stages

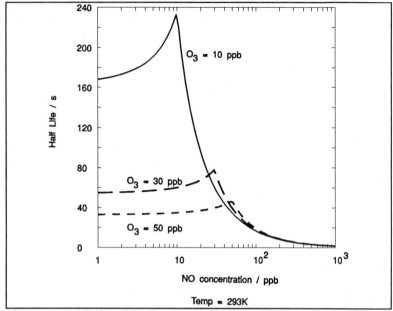

Figure 1 The half life for the reaction between NO and O_3.

of dispersion the relative rate of turbulent dispersion is very fast and it is likely that chemical reaction between the mixing species will be limited by the small scale mixing. Further downwind small scale gradients within the plume are reduced with respect to the overall plume size, and the overall rate of NO oxidation becomes limited by the rate of turbulent mixing. Hegg et al (1977) estimate the ratio of chemical to dispersion timescale for a set of plume observations and conclude that NO conversion was generally turbulent dispersion controlled out to the distances measured in their study (90 km) and travel times up to 4 h. Cheng et al (1986) show a linear dependence (albeit with considerable scatter) between NO conversion rate and the one third power of the turbulence dissipation rate, again suggesting the controlling influence of turbulent dispersion. An extreme case, which will be discussed further below, is reported by Cocks et al (1983). This shows clear limitation of O_3 destruction in a very poorly dispersed plume at a distance of 650 km from the source.

During daylight hours an equally important process to the O_3 destruction described above is the photolysis of NO_2 by uv radiation:

$$NO_2 + h\nu\,(\lambda < 430\,nm) \rightarrow NO + O(^3P) \qquad (3)$$
$$O(^3P) + O_2 \rightarrow NO_2$$

The second step is very fast and rate of O_3 formation is determined by the first step. The rate at which this occurs is denoted J_{NO2} and depends upon the quantum yield, absorption cross section and total radiative flux from all directions as a function of wavelength. The first two of these can be measured in the laboratory, but the last depends not only upon the solar spectrum but upon the scattering properties of the atmosphere.

In general, in order to calculate the total solar flux, and hence J_{NO2}, a radiative transport model must be used which takes account of the atmospheric distribution of absorbing and scattering materials such as O_3, aerosol particles and cloud. Derwent and Hov (1979 and numerous subsequent publications) use a simple exponential function of the form

$$J_{NO_2} = a \exp(b/\cos\theta) \qquad (4)$$

where θ is the solar zenith angle and a and b are constants obtained by fitting to the results of a relatively simple horizontally homogeneous two stream radiative transfer model of Isaksen et al (1977). This fit has been used to derive Fig. 2, which shows the half life of NO_2 with respect to photolysis at various latitudes.

On the whole, in sunny conditions, the half life is of order a minute, but this can be considerably increased under cloud cover or when the sun is low in the sky. To make proper allowance for this the cloud cover (and any anomalous conditions such as high aerosol loading) throughout the atmosphere must be included in a radiative transport model together with knowledge of the local surface albedo. Demerjian et al (1980) present results from such a model, together

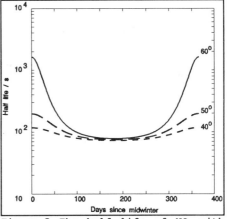

Figure 2 The half life of NO_2 with respect to photolysis, calculated for local noon using the model of Derwent (1981).

with comparisons against a J_{NO2} measuring device which detects the NO produced in a reaction chamber designed to capture light equally from all directions. Agreement between the model and observations is remarkably good. However, this is at the expense of a requirement for input data about atmospheric conditions which are unlikely to be available routinely.

An alternative approach is to monitor the radiation field

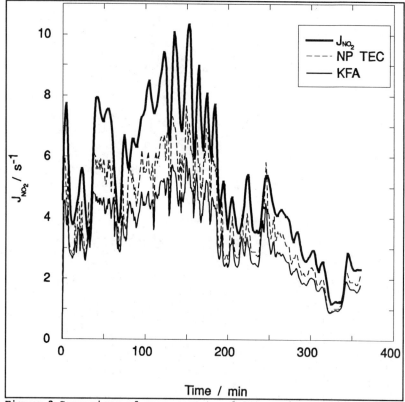

Figure 3 Comparison of measurements of JNO_2, using a spectrometer, the KFA JNO_2 instrument and the NP TEC dodecahedron. The latter two are uncalibrated signals.

directly. Fig. 3 shows a comparison between three monitoring systems performed at NPTEC (R. Wright, Personal communication). The 'primary measurement' comes from an Oriel diode array specrometer with 0.5 nm resolution which analyses light from a quartz collector designed to collect light equally efficiently from all directions. The spectra so obtained are then combined with measurements of quantum yield and cross section from the literature to obtain J_{NO2}. The other two devices make direct measurements of the integrated flux from all directions over a waveband relevant to NO_2 photolysis. The first is a device using a single detector and a light collector for each hemisphere, designed by KFA Jülich, while the second (NPTEC) device comprises a dodecahedron with a photodiode and filter on each face apart from the supporting one. The figure shows only raw signals from the latter two devices as the overall calibration is uncertain, but it is clear that both are capable of reproducing the detailed structure of the signal under differing cloud conditions and zenith angles.

The Photostationary State

If we consider a system comprising just NO, NO_2 and O_3 in air, at concentrations at which eq. (1) is negligible, then the concentrations of these species tend to the well known photostationary state (PSS), in which the following relationship holds:

$$\frac{[NO][O_3]}{[NO_2]} = \frac{J_{NO_2}}{k_2} \qquad (5)$$

In this simple system both the total NOx (NO + NO_2) and the sum of O_3 and NO_2 are conserved quantities. The first obviously derives from the source(s) in question, while the latter has been dubbed 'oxidant' (White, 1977). Given the PSS and these two conservation relations it is straightforward to derive the concentrations of NO, NO_2 and O_3. If we denote the oxidant by Ox, then we obtain:

$$NO_2 = \frac{(NO_x + Ox + J_{NO_2}/k_2) \pm \sqrt{((NO_x + Ox + J_{NO_2}/k_2)^2 - 4NO_x \cdot Ox)}}{2} \qquad (6)$$

The sign chosen depends upon whether NOx exceeds Ox.

Since NOx and Ox are conserved quantities in the system we expect them to be transported like any passive tracer and can therefore apply normal dispersion models to calculate their transport. If the PSS applies we can use eq. (6) to estimate NO_2 levels given ambient O_3 mixing into the plume. It is therefore of considerable interest to know under what conditions the PSS is observed to hold and the general magnitude of deviations from it.

One problem arises when interpreting measurements. Most studies of very nearfield plumes have been performed by traversing the plume using aircraft. Unfortunately the response time of most gas instruments is insufficient to resolve the small scale fluctuations and some degree of spatial averaging occurs. Janssen and Elshout (1987) illustrate this by comparing the signal from a fast response temperature sensor, which shows a very narrow plume 0.5 km from the source, to the signals from NOx, SO_2 and O_3 detectors, which all show the effect of finite time response. Of particular importance is the apparent presence of O_3 in the plume which is largely an instrumental artefact.

Bilger (1978) considers the effect of fluctuations not resolved by measurement instruments and molecular diffusivity. He concludes that, except under unusual conditions, the PSS ratio measured will generally exceed the PSS constant even if PSS is attained on a point by point basis, and that the latter is difficult to achieve as a result of molecular diffusion. Janssen and Elshout (1977) show substantial positive deviations from the ratio out to around 6 km from a source. White (1977) describes a very elegant set of data illustrating well the features described above, namely the conservation of oxidant and the PSS. In contrast Hegg et al (1977) find substantial positive deviation out to 40 km, although there is a decreasing trend to the deviation with distance.

Given the difficulty of making short range measurements it is of interest to consider the results of models. Carmicheal and Peters (1981) propose a simple semi-empirical model of inhomogeneous mixing and, after fitting to data originally reported by Davis et al (1974) from aircraft traverses of the Morgantown power plant in Maryland, conclude that substantial positive deviations from PSS occur within 5-10 minutes of the source. This agrees with the calculations of Varey et al (1978), using a diffusion model.

Modelling the Ground Level Impact of NOx emissions

The degree of mixing (and hence the time) required before PSS is approximately attained corresponds roughly to the degree of mixing required before typical plumes reach the ground (although there is no mechanistic connection). From the point of view of ground level impact it may therefore be sufficient to assume that PSS applies. As noted above this will generally over-estimate the formation of NO_2 to some extent, but the over-estimation is likely to be small by the time the plume reaches the ground.

Bange et al (1991) show a comparison between two models and observations of NO_2 formation in Dutch power plant plumes made during more than 60 measuring flights up to 20 km from the source. The first model is a diffusion model incorporating plume chemistry while the second is a simple Gaussian plume for NOx followed by the assumption of PSS. They find that, with the important proviso that instantaneous plume dimensions are used rather than time averaged dimensions, the PSS model is perfectly adequate for predicting the proportion of plume NOx oxidised to NO_2 under most circumstances. This appears to be the case even though the PSS is not strictly attained because in the very nearfield the NOx concentration generally greatly exceeds the ambient O_3 concentration, so that within the plume the O_3 concentration is reduced to a small value even if it does not reach the PSS value. Deviation from the PSS tends, in relative terms, to reveal itself primarily, therefore, in the O_3 concentration rather than the NO_2 concentration.

Using the simple model of Derwent and Hov (1979) the PSS ratio has been calculated (for noon) and is shown in Fig. 4. At European latitudes the PSS constant is around 20 ppb under clear sky, high sun conditions and most of the

time will be rather less.
Under conditions that are
likely   to   be   of
environmental   concern
(from the point of view of
NO₂ impact from a local
s o u r c e)   t h e   N O x
concentration in a plume
is likely to exceed this
by a considerable factor.

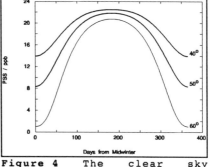

A simple extension to the
PSS   approach   when
estimating ground level NO₂
impact when information
regarding the radiative
flux is not available is

Figure 4   The   clear   sky
Photostationary State Constant at
Noon as a function of time of year
and latitude.

to assume that J_{NO2}/k is zero, which implies that either
NO or O_3 are zero. In this case NO_2 produced by reaction
with O_3 is simply given by the minimum of total NOx or
ambient O_3, an assumption which which will yield
conservative results.

This (and the PSS) assumption only applies to the
instantaneous plume, so care must be used when applying
the conventional dispersion models which generally
compute hourly mean concentrations. The resulting NO_2
must be reduced to allow for the proportion of each hour
that the instantaneous plume is not present at any give
point (the intermittency). This is typically about two
thirds, so that the hourly mean NO_2 is limited to about
one third the ambient O_3, but clearly the particular
dispersion conditions, in particular stability, must be
taken into account (Bange et al, 1991).

3 PHOTOCHEMICAL OXIDANT GENERATION IN PLUMES

From the point of view of photochemical oxidant
generation the PSS has essentially no impact, as it
simply results in the interconversion of O_3 and NO_2. In
order to generate oxidant a mechanism is required which
converts NO to NO_2 without consuming O_3. The termolecular
reaction discussed above is one such mechanism, but it
tends to be of minor importance as a result of dilution
quenching. It now well established that the primary
meachanism for converting NO to NO_2 in the atmosphere is
reaction with peroxy radicals:

$$NO + RO_2 \rightarrow NO_2 + RO$$
$$NO + HO_2 \rightarrow NO_2 + OH \tag{7}$$

The organic peroxyradicals arise from the abstraction of
a hydrogen atom followed by the addition of oxygen:

$$RH + OH \rightarrow R + H_2O$$
$$R + O_2 \rightarrow RO_2$$

(8)

Since large combustion plant contain very little VOC it is clear that a pre-requisite for PCO formation is the mixing of VOCs into the plume from outside. To this extent it may be said that the VOCs 'cause' any PCO generation which might occur in the plume, although, of course, both VOCs and NOx are required.

The presence and rate of PCO generation is critically dependent, amongst other things, upon the relative concentrations of NOx and VOC, and hence not only upon plume and ambient emissions but also upon the rate of intermixing. It is reasonable to suppose that faster mixing, higher ambient VOC levels (in relation to NOx) and higher u.v. radiation levels will all promote PCO generation. Given the number of different variables of importance and the range over which they all may vary it is perhaps not surprising that a great variety of observations exist in the literature.

Many observations have been reported which show O_3 levels in plumes less than or equal to those in surrounding air. It should be noted that this, in itself, does not imply that oxidant has not been generated in the plume, only that any generation inside the plume has been less than in the surrounding air. As stated above, evidence for oxidant generation comes from NO_2 levels in a plume greater than the apparent deficit of O_3.

White (1977) shows aircraft measurements on 14 August 1974 out to 45 km (3-4 h) from the Labadie power plant, all with O_3 depression. No evidence was found of extra O_3; instead there is even some suggestion of reduced O_3 at 45 km possibly as a result of plume scattering and hence reduced photolysis rates inside the plume. Similarly, Melo and Stevens (1981) show a 1:1 relationship between NO_2 formed and O_3 consumed from the Nanticoke and Lakeview (on the shores of the Great Lakes) plume study. No O_3 production was observed out to 93 km from the source in November 1975, and 66 km October 76. In June and Sept similar results were found at shorter range, but unfortunately no measurements were made at longer range.

Hegg et al (1977) report measurements on two coal and two gas-fired power plants, the coal plants being in Centralia, Washington and Farmington, New Mexico and the gas at Hobbs, New Mexico and Longview, Texas. The measurements are unusual in that they include results from an Epply uv radiometer. O_3 levels greater than ambient were never seen up to 90 km, or 4 h from the source. High O_3 (~120 ppb) was observed on three occasions but detailed analysis revealed this resulted from mixing of high levels from above the boundary layer as the boundary layer inversion broke up.

Richards et al (1981) report measurements in June, July and December 1979 on the plume from the Navajo Generating Station at Page, Arizona. They again found that the ratio of NO_2 to O_3 was consistent with the PSS, and that there was no evidence of O_3 production out to their maximum range of 115 km from the source.

Other authors, however, have reported clear O_3 production at distances of around 100 km from the source. For example Forrest et al (1981) observed small O_3 plumes on distant traverses and cite an example 107 km from the Cumberland Plant, TVA, in August 1978 of a 25 ppb peak on top of 25 ppb ambient O_3. As part of the same study Gillani et al (1981) show measurements at 110 and 160 km from the Cumberland plant. The measurements at 110 km show the phenomenon of O_3 'wings', i.e. elevated O_3 at the plume edges while O_3 remains depressed at the plume centre. This clearly indicates the importance of the decreasing ratio of NOx to VOC as dispersion proceeds, although the phenomenon itself had disappeared at 160 km to be replaced by a fully developed O_3 plume.

Gillani et al (1978) emphasise the importance of the composition of the air into which a plume mixes and report a substantial O_3 plume on July 9 1976 in the Labadie plume 190 km (9h) downwind (the same traverse also shows a clear O_3 deficit in the Kincaid plume 12 km downwind as it mixes into an O_3 plume resulting from the to St. Louis urban area. In contrast July 18 was more active, both in terms of turbulent mixing (resulting from greater instability) and ambient O_3. On this occasion there is evidence of some ozone production after only 30 km.

Davis et al (1974) report some of the first measurements of O_3 production, and also some of the closest to the source. Some evidence of an O_3 'wing' is shown as close as 24 km downwind and by 40 km this had developed into a clear O_3 plume, peaking at 100 ppb compared with the (high) ambient levels of around 80 ppb.

Meagher et al (1981b) report several helicopter borne plume measurements. In 13 out of 14 cases the NO_2 produced was less than the O_3, the difference being attributed to the loss of NO_2 by reaction with OH radicals to form nitric acid. One observation, however, on 23 August (1978) shows O_3 production beyond 50 km. This was attributed to mixing of the plume with polluted air from Chattanooga, Tennessee.

All of the above observations of O_3 production were made in North America. No similar reports exist relating to European sources. Indeed the most extreme counter example is shown by Cocks et al (1983) who report a clear O_3 deficit 650 km from the source (a group of three 2000 MW plant). The observation was made just off the Danish coast after the plume had crossed the North Sea from the UK. The observations were made in midwinter, in a cloud-filled maritime boundary layer in which lateral dispersion had all but ceased. They are thus, in themselves, unsurprising but do serve as a reminder that NO emissions can still have a negative effect on PCOs at long range. Much more photochemically active conditions were encountered by the same group studying the same sources in June 1981 , Fig. 5, (PORG I). On this occasion some depression of O_3 was visible (though with possible 'wings') at the longest distance studied

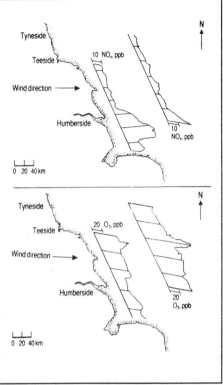

Figure 5 Measurements of NOx and O_3 downwind of central and northern England during 18 June 1980.

(180km), even though a very marked O_3 bulge had developed in the air that had passed over the Teeside area, an urban area with major industrial and oil refinery sources.

In an effort to study conditions more conducive to PCO production, namely the interaction of power plant plumes with urban emissions, numerous airborne studies have been performed around the London area including the Thameside power stations to the east of London. None have shown O_3 production in a plume above that inside the urban airmass, although several do show the generation of oxidant inside the plume. Fig. 6 shows results from traverses made on two separate flights, on the morning and afternoon of 12 July 1988. The first traverse was performed just downwind of London and shows the characteristic drop in O_3 in the power station plume. A similar drop (around 7-8 ppb) occurred in the total oxidant, showing that photochemical activity was suppressed in the plume. The second traverse was performed in approximately the same air somewhat further downwind, at a time approaching the typical time that peak O_3 levels tend to be observed. An O_3 dip is still visible in the power station plume, but much smaller (only about 2 ppb). Total oxidant is still rather less than that in the urban plume (around 6ppb), but the absolute level exceeds the level observed in the surrounding air in the morning, demonstrating that additional oxidant was present in both the plume and urban air. It is not possible, without using numerical modelling, to establish how much of the oxidant in the plume was generated *in situ* and how much had mixed in from the surroundings.

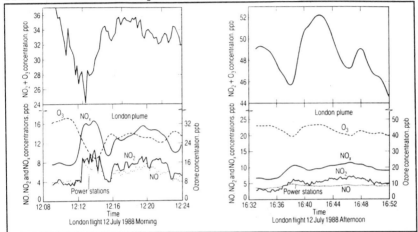

Figure 6 Observations downwind of London on the morning of 12 July 1988 and in the same air during the afternoon.

Given that photochemical activity is less inside the plume than outside it might be expected that the concentration of other oxidising species such as OH and HO_2 radicals and H_2O_2 will also be less. H_2O_2 is a particularly sensitive indicator as its production rate depends upon the square of the HO_2 concentration. Fig. 7 shows an interesting observation of gas phase H_2O_2 during repeated traverses of the London plume and Thameside power stations. A clear deficit in H_2O_2 was observed inside the power plant plume. No cloud was present to act as a sink for H_2O_2 by reaction with SO_2 inside the plume, and it appears instead that H_2O_2 production was actually up to 50% greater outside the plume than inside.

4 NUMERICAL MODELLING

It would appear from observations that PCO generation inside power plant plumes around 100 km from the source is a relatively frequent event, at least in summer, over North America, while excess PCO generation in European plumes occurs rarely if ever. Since the laws of physics and chemistry are (presumably) the same on both sides of the Atlantic it seems likely that the explanation for this difference lies in differences in either or both the composition of ambient air and the meteorological

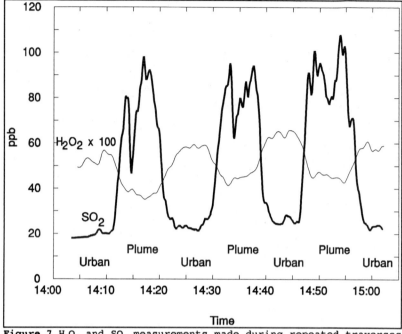

Figure 7 H_2O_2 and SO_2 measurements made during repeated traverses of a power station plume dispersing into the London plume.

conditions. Numerical models can help to explain these differences and, by extrapolation, can be used to indicate the likely changes that would result from a change in emissions.

It is clear from the preceeding discussion that any power plant plume chemistry modelling must take into account the chemistry of the surrounding air and pay particular attention to the rate of mixing between the two. The complexity of the chemical reaction schemes generally required precludes a very detailed turbulence model, and instead the common approach is based upon the single or multiple expanding box model.

Stewart and Liu (1981) model Widows Creek and Oak Creek power plants using an expanding plume model which assumes homogeneous mixing in the vertical while lateral mixing is treated by modelling the plume as a number of cells or boxes which are allowed to move and expand downwind in a prescribed manner. They report modest increases in the cross plume average O_3 as a function of distance downwind beyond about 33 km for Widows Creek source (August 1978) and 50 km for Oak Creek source (July 1977). In both cases the ambient air exhibited a substantial growth in O_3 over the same period, indicating that the air into which the plume was mixing was already photochemically active. Both cases involved the mixing with moderately urban and rural air but in neither case were measurements of VOCs in the ambient air available. Instead 'typical' concentrations were assumed. In spite of this agreement between the model and observations appears to be quite good.

Hov and Isaksen (1981) use a similar model to simulate the Labadie plume in St Louis. The simulation shows O_3 bulges after 2.5-3 h travel (or more), 10-20% above ambient levels. Prior to the formation of a bulge the presence of small 'wings' is predicted as the material in the outside of the plume is more dilute. On the whole the model is able to reproduce the observed behaviour of the Labadie plume, mentioned above, relatively well. Of particular interest are a series of model 'experiments', in which various parameters are changed to reproduce the gross effects of different scenarios. The results of these may be summed up as follows:

1) Ozone production is generally favoured by increased mixing with surrounding air.

2) Solar radiation must be reduced substantially before there is any major effect on secondary pollutants. A 40% reduction (by cloud) leads to only a small decrease in excess O_3, and a 60% reduction is required before the excess disappears.

3) Initial plume concentrations have a marked but very non-linear impact on excess O_3. A doubling leads to only about 25% increase in excess O_3 after 8 h travel, while a further doubling reduces the excess by a factor of three.

The size of the emission source, in relation to ambient emissions, is thus seen to be of crucial importance. In agreement with the observations reported above when 'typical' UK emissions and solar radiation are used no excess O_3 results. Similar conclusions are drawn by Cocks and Fletcher (1988 and 1989). They use a physically simpler, single expanding box model (which, according to Clark and Cocks (1988), yields similar results to a multiple box model at long range providing appropriate dispersion parameters are chosen). In the case of the second paper a second 'box' is introduced to represent the rural emissions into which the plume is dispersing and which, itself, disperses into a 'rural' airmass.

Cocks and Fletcher find that plume behaviour depends in a complex way on a number of factors. The time of year is clearly of importance as it determines the overall level of photochemical activity. The time of emission is also found to be important as it determines the degree of dispersion which can occur, and hence the admixture of outside air containing VOCs and other species, before the action of sunlight. The dispersion rate has an important

impact on oxidant levels in the plume and the rate of
plume development, although the dispersion model used may
have some difficulty in accurately simulating situations
in which a plume is released at night above the boundary
layer and disperses very little before morning when a
developing convective boundary layer leads to fumigation.

In common with Hov and Isaksen (1981), but over a much
wider range of conditions, Cocks and Fletcher find that
ozone, H_2O_2 and PAN levels in the plume are generally
lower than in surrounding air even after 24h, and only in
relatively slowly dispersing plumes is ozone production
at all significant after 24h, and even then less is
produced than in the surrounding air. It is particularly
noteworthy that the prediction regarding H_2O_2 production
appears to be confirmed by the observations shown in Fig.
7 above.

5 CONCLUSIONS

The results discussed above may be summarised by the
following conclusions.

The Very Near Field

1) In the very near field, before the plume has reached
the ground, plumes do not generally reach the
photostationary state as the rate of turbulent mixing is
faster than the rate of reaction.

2) In general there will be less NO_2 than the
photostationary state would predict.

3) The termolecular reaction between NO and O_2 may
contribute significantly to the production of NO_2,
especially in poorly dispersing plumes.

The Near Field

1) In the near field NO_2 appears to be well predicted by
the photostationary state.

2) A conservative, and reasonably accurate, assumption
would be that NO_2 is given by the minimum of ambient
ozone and plume NOx.

3) When computing hourly means, the plume intermittency
must be taken into account. This reduces the hourly mean
concentration of NO_2 considerably, typically by a factor
of three.

The Mid Field.

1) American data show regular occurrence in summer sunny
conditions of plume ozone greater than ambient at

distances greater than approximately 100 km. Occasionally this may occur as close as 40-50 km.

2) UK data show no similar effects: ozone is always depleted at this distance.

3) These observations are broadly, and often quite accurately, confirmed by model simulations. The differences between the UK and North America are many, but the most important are probably different emissions (a higher NOx/VOC ratio and lower natural VOCs in the UK), differences in mixing (greater frequency of convective conditions) and differences in upwind air.

The Far Field

Models suggest that in midsummer, under favourable conditions the ozone in plumes may start to exceed ozone in surrounding air after about a day, but the results are critically dependant upon many factors, notably dispersion rate and intervening sources.

6 ACKNOWLEDGEMENTS

The author very gratefully acknowledges the contribution of many colleagues to this paper. In particular he thanks Dr. Raymond Wright for permission to use his photolysis data (Fig. 3) and Dr. Tony Marsh and Dr. Paul Lightman for their aircraft data (Fig. 7).

7 REFERENCES

Atkinson, R. and Lloyd, A.C., 1984, Evaluation of kinetic and mechanistic data for modelling of photochemical smog, J. Phys. Chem. Ref. Data **13**, 315-444.

Bange, P., Janssen, L.H.J.M., Nieuwstadt, F.T.M., Visser, H. and Erbrink, J.J., Improvement of the modelling of daytime nitrogen oxide oxidation in plumes by using instantaneous plume dispersion parameters, Atmos. Env. **25A**, 2321-2328.

Baulch, D.L., Cox, R.A., Hampson, R.F., Kerr, J.A., Troe, J. and Watson, R.J., 1984, Evaluated kinetic and photochemical data for atmospheric chemistry, J. Phys. Chem. Ref. Data **13**, 1259-1375.

Bilger, R.W., 1978, The effect of admixing fresh emissions on the photostationary state relationship in photochemical smog, Atmos. Env. **12**, 1109-1118.

Carmichael, G.R. and Peters, L.K., 1981, Application of the mixing-reaction in series model to $NOx-O_3$ plume chemistry, Atmos. Env. **15**, 1069-1074.

Cheng, L., Peake, E., Rogers, D. and Davis, A., 1986, Oxidation of nitric oxide controlled by turbulent mixing in plumes from oil sands extraction plants, Atmos. Env. **20**, 1697-1703.

Clark, P.A. and Cocks A.T., 1988, Mixing models for the simulations of plume interactions with ambient air. Atmos. Env. **22**, 1097-1106.

Cocks, A.T., Kallend, A.S. and Marsh, A.R.W., 1983, Dispersion limitations of oxidation in power plant plumes during long-range transport, Nature **305**, 122-123.

Cocks, A.T. and Fletcher, I.S., 1988, Major factors influencing gas-phase chemistry in power plant plumes during long-range transport - I. Release time and dispersion rate for dispersion into a 'rural' ambient atmosphere, Atmos. Env. **22**, 663-676.

Cocks, A.T. and Fletcher, I.S., 1989, Major factors influencing gas-phase chemistry in power plant plumes during long-range transport - II. Release time and dispersion rate for dispersion into a 'urban' ambient atmosphere, Atmos. Env. **23**, 2801-2812.

Davis, D.D., Smith, G. and Klauber, G., 1974, Trace gas analysis of power plant plumes via aircraft measurement: O_3, NOx and SO_2 chemistry, Science **186**, 733-736.

Demerjian, K.L., Schere, K.L. and Peterson, J.T., 1980, Theoretical estimates of actinic (Spherically integrated) flux and photolytic rate constants of atmospheric species in the lower troposphere, Adv. Env. Sci Tech. **10**, 369-459

DeMore, W.B., Golden, D.M., Hampson, R.F., Howard, C.J., Kurylo, M.J., Molina, M.J., Ravishankara, A.R. and Sander, S.P., 1987, Chemical kinetics and photochemical data for use in stratospheric modeling, JPL Publication 87-41, 1.

Derwent, R.G. and Hov, O, 1979, Computer modelling studies of photochemical air pollution formation in North West Europe, UKAEA Report AERE-R 9434.

Forrest, J., Garber, R.W. and Newman, L., 1981, Conversion rates in power plant plumes based on filter pack data: the coal-fired Cumberland plume, Atmos. Env. **15**, 2273-2282.

Gillani, N.V., Husar, R.B., Husar, J.D., Patterson, D.E. and Wilson, W.E., 1978, Project MISTT: kinetics of particulate sulphur formation in a power plant plume out to 300 km, Atmos. Env. **12**, 589-598.

Gillani, N.V., Kohli, S. and Wilson, W.E., 1981, Gas-to-particle conversion of sulphur in power plant plumes-I. Parameterization of the conversion rate for dry, moderately polluted ambient conditions, Atmos. Env. **15**, 2293-2313.

Hampson, R.F., Jr. and Garvin, D., 1977, Reaction rate and photochemical data for atmospheric chemistry - 1977, NBS Spec. Publ. 513, May 1978.

Hegg, D., Hobbs, P.V., Radke, L.F. and Harrison, H., 1977, Reactions of ozone and nitrogen oxides in power plant plumes, Atmos. Env. **11**, 521-526.

Hov, O and Isaksen, I.S.A., 1981, Generation of secondary pollutants in a power plant plume: a model study, Atmos. Env., **15**, 2367-2376.

Isaksen, I.S.A, Midtbo, K.H., Sunek, J. and Crutzen, P.J., 1977, A simplified method to include molecular scattering and reflection in calculations of photon fluxes and photodissociation rates, Geophysica Norvegica, **31**, 11-26.

Janssen, L.H.J.M. and Elshout, A.J., 1987, Formation of NO_2 in power-plant plumes: measurement and modelling, KEMA Sci. & Tech. Reps **5**, 259-279.

Janssen, L.H.J.M., 1988, Reactions of Nitrogen Oxides in Power-Plant Plumes, Doctoral Thesis, Delft NL.

Meagher, J.F., Stockburger III, L., Bonanno, R.J. and Luria, M., 1981a, Cross-sectional studies of plumes from a partially SO_2 -scrubbed power plant, Atmos. Env. **15**, 2263-2272.

Meagher, J.F., Stockburger III, L., Bonanno, R.J., Bailey, E.M. and Luria, M., 1981b, Atmospheric oxidation of flue gases from coal fired power plants - a comparison between conventional and scrubbed plumes, Atmos. Env. **15**, 749-762.

Melo, O.T. and Stevens, R.D.S., 1981, The Occurrence and nature of brown plumes in Ontario, Atmos. Env. **15**, 2521-2529.

Miller D.F., Alkezweeny A.J., Hales J.M. and Lee R.N., 1978, Ozone formation related to power plane emissions. Science **202**, 1186-1190.

Richards, W.L., Anderson, J.A., Blumenthal, D.L., Brandt, A.A., McDonald, J.A. and Waters, N., 1981, The chemistry, aerosol physics, and optical properties of a western coal fired power plant plume. Atmos. Env. **15**, 2111-2134.

Romer, F.G., van Duuren, H., Elshout, A.J. and Viljeer, J.W., 1979, Measurements from aircraft of the propagation and conversion of primary air-pollution components in smoke plumes, VGB Conference May 1979, "Power Stations and the Environment".

Stewart, D. and Liu, M., 1981, Development and application of a reactive plume model, Atmos.Env. **15**, 2377-2393.

Varey, R.H, Sutton, S. and Marsh, A.R.W., 1978, The oxidation of nitric oxide in power station plumes - a numerical model, CERL Laboratory Report, RD/L/N 184/78.

White, W.H., 1977, NOx-O_3 Photochemistry in power plant plumes: comparison of theory with observation, Environmental Sci. & Tech., **11**, 995-1000.

Measurements of Urban Photochemical Oxidants

J. S. Bower, G. F. J. Broughton, and P. G. Willis

WARREN SPRING LABORATORY, GUNNELS WOOD ROAD, STEVENAGE,
HERTFORDSHIRE SGI 2BX, UK

SUMMARY

Recent measurements of ozone and oxides of nitrogen in urban areas of the UK
are reviewed. For this purpose, results from two representative monitoring locations
in London are primarily considered - the long-running background Central London
and kerbside Cromwell Road stations.

Baseline ozone concentrations at a range of UK monitoring stations are analysed
in terms of chemical and physical loss mechanisms, with the NO_x sink shown to be
the dominant determining process in urban areas.

Ambient concentrations of NO_2, NO_x and O_3 are strongly interrelated in urban areas,
and their dependence on meteorological and dispersion conditions, emissions and
proximity to primary pollutant sources is investigated.

A variety of analyses demonstrate a non-proportional relationship between NO_2 and
NO_x concentrations over a variety of timescales. This has important implications
for future NO_x emission control strategies. These analyses also demonstrate that
elevated NO_2 levels in kerbside environments, although not wholly explainable on
the basis of primary vehicle emissions alone, can be produced by the fast titration
reaction with ambient ozone.

1   INTRODUCTION

Motor vehicles and stationary source fuel combustion - primarily in power stations and industry - represent the major anthropogenic emissions of nitrogen oxides (NO_x = $NO + NO_2$) in the UK. Typically, the NO_2 content of combustion emissions is ~ 5 - 10% of the total NO_x. Stationary and vehicular sources currently contribute approximately 40 and 48% respectively of total man-made NO_x emissions (Leech, 1991). Whilst total stationary source emissions have remained fairly stable since 1970, those from transport have increased steadily over this period.

Because of their proximity and low emission heights, vehicle pollutant emissions are of particular significance for urban air quality, and can account for 70% of NO_2 concentrations in urban areas and up to 90% in episode conditions (Simpson, 1987, Simpson et al, 1988).

Ambient concentrations of NO_2, NO_x and O_3 are strongly interrelated. The relevant chemistry is well known, so only a brief summary is here provided.

In the absence of local pollutant emissions and of photochemically generated ozone, an equilibrium is quickly established (in a few minutes typically in a well-mixed air mass) between NO, NO_2 and ozone.

The photo-dissociation of NO_2 ($\lambda \leq 420$ nm):
$$NO_2 \rightarrow NO + O \tag{i}$$

together with the subsequent reaction:
$$O + O_2 + M \rightarrow O_3 + M \tag{ii}$$

and ozone scavenging by NO -
$$O_3 + NO \rightarrow NO_2 + O_2 \tag{iii}$$

result in the well-known photostationary equilibrium between these species:
$$O_3 + NO \rightleftharpoons NO_2 + O_2 \tag{iv}$$

In irradiated polluted atmospheres, however, this equilibrium state is perturbed: peroxy radicals formed by the photochemical degradation of hydrocarbons react with NO to form NO_2, without the direct involvement and consequent loss of ozone (reaction iii):
$$RH + OH \rightarrow R* + H_2O \tag{v}$$

$$R* + O_2 \rightarrow RO_2 \tag{vi}$$

$$RO_2 + NO \rightarrow RO + NO_2 \tag{vii}$$

The net result of these reactions is to shift equilibrium reactions (i - iii) during the daytime hours towards the net formation of NO_2 and then O_3.

In addition to reaction (iv) and (v) - (vii), it may be noted that NO_2 can also be produced by reaction of NO with O_2, although this process is relatively slow at typical ambient concentrations:

$$2NO + O_2 \rightarrow 2NO_2 \qquad\qquad \text{(viii)}$$

The major chemical sinks for NO_2 include reaction with OH radicals to form HNO_3 (during the day) and with O_3 (by night). Removal of ozone from the atmosphere results from dry deposition to the ground, reaction with unsaturated hydrocarbons and - by night - scavenging by NO_2 or NO.

The UK national monitoring networks for nitrogen oxides and ozone have expanded significantly since 1987. In particular, extended national networks for NO_2 and O_3, covering a variety of rural and urban location types, are currently managed or operated by WSL for Department of the Environment (Lampert, 1991). In this paper, the focus is on urban measurements. For this purpose, results from two long-running representative monitoring locations in Central London are reviewed and analysed. The emphasis is not on documenting measurements, which have been widely reported elsewhere (Bower et al, 1991(i) - iii), Bower et al 1989, i), Broughton et al, 1992), but on exploring the relationship between NO_2, NO_x and O_3 concentrations in urban environments.

2 EXPERIMENTAL

Data from the Central London (CLL) and Cromwell Road (CRD) monitoring station are primarily considered here, although results from corresponding national network sites in other parts of the country are assessed where appropriate. The Central London Laboratory site, operational since 1972, (Grid Ref TQ 292791) is located in a residential/commercial area in Victoria and is well placed to provide background urban air quality measurements of NO_x and O_3. By contrast, the Cromwell Road site (Grid Ref TQ 264789), operational since 1973, is situated at the kerbside of a major road with a traffic throughput of 50 - 60,000 vehicles per day. NO_x measurements from here are therefore indicative of near-source conditions.

Chemiluminescent analysers at these sites measure NO and NO_x levels, with NO_2 derived from the difference between these concentrations. A UV absorption analyser provides continuous O_3 measurements at Central London. The CRD and CLL stations, together with other national monitoring stations operated or managed by WSL for DoE, are highly automated: an intelligent data logger records spot concentrations every 10 seconds and accumulates these to provide 15-minute averages. The logger also monitors instrument alarm and diagnostic functions and controls daily instrument zero/span response checks. These are implemented using a separate NO_2 permeation tube oven and zero air generator (for NO_x analysers) and internal zero/span modules for O_3 analysers. NO_x analysers are also audit calibrated at weekly intervals using commercially-sourced NO in N_2 and NO_2 in air cylinders, carefully cross-referenced against primary gas standards (permeation tube, static dilution and gas phase titration-derived). Ozone analysers are also manually audited every week.

Monitoring sites are linked to a data telemetry system which allows the transmission of results, over standard telephone lines, to a computerised network control centre at Stevenage. Data are acquired automatically twice daily, although hourly polling is initiated during high pollution episodes. Results are subsequently validated and processed using a microcomputer-based system controlled prior to databasing on a pc local area network and the main laboratory workstation network. Data are also rapidly transferred (within 1 hour) by direct terminal link to the

Meteorological Office for subsequent public dissemination via press and TV, as well as to other on-line data users within central and local government.

Comprehensive quality assurance and control (qa/qc) procedures have been developed and implemented by WSL for site operation, analyser calibration and data processing. These procedures allow data quality and capture rates to be maximised and ensure that network objectives are fulfilled. Full information on this qa/qc programme is available elsewhere (Bower et al, 1989, ii).

Even with the adoption of stringent qa/qc methodologies, the accuracy and precision of measurements remain important factors in assessing the validity of results presented in this paper. Detailed laboratory and field evaluations of ambient analysers, together with a full assessment of operation/calibration practice have demonstrated that O_3 data reported here will be accurate to within $\pm$ 11%, with an associated precision of $\pm$ 2 ppb. Corresponding results for NO are (A = $\pm$ 10%, P = $\pm$ 2.5 ppb) and for NO_2 (A = $\pm$ 10 - 11%, P = $\pm$ 3.5 ppb).

Network intercalibration exercises are performed every six months using a specially designed photometer system for O_3 and transfer standard NO/NO_2 sources prepared at WSL's national gas calibration centre. These intercalibrations ensure full intercompability of data from all measurement sites and provide direct traceability to absolute or national metrology standards. Interlaboratory calibration exercises have also demonstrated excellent agreement between WSL primary gas standards and those maintained by AEA Harwell (UK), NOAA (United States), RIVM (Holland) and UBA (Germany) (Sweeney, 1991).

3   SHORT-TERM O_3/NO_x VARIATIONS IN URBAN AREAS

The strong interaction between O_3, NO and NO_2 in urban areas is readily apparent when examining diurnal variations of these pollutants at Central London (Fig 1a - c). These are graphed against local time to highlight man-made influences on ambient concentrations, whilst separate graphs of summer and winter periods demonstrate seasonal differences. Corresponding weekday and weekend analyses highlight the impact of primary NO_x emissions from motor vehicles and other sources.

The NO_x sink for ozone is likely to be the dominant loss process for this pollutant wherever local NO concentrations are comparable to the 'background' boundary layer O_3 concentration of approximately 35 ppb. Measurements carried out by WSL (Williams et al, 1988) suggest that this will be the case in most reasonably sized urban areas of the UK. In a six-month study at over 400 urban locations in 1986 (Bower et al, 1991, iii)), 90% of the non-kerbside measurement sites had average NO_2 concentrations exceeding 10 ppb and may be expected to have corresponding NO concentrations well over 20 - 30 ppb.

Since the major interactions between NO, NO_2 and O_3 in urban areas are well understood, only a brief qualitative overview is here provided.

The strong influence of primary NO_x emissions during traffic rush hour periods is evident during both summer and winter periods. During the summer months (Fig 1a), primary NO emitted during the morning is scavenged quickly by ozone and - more slowly - by peroxy radicals formed by OH radical attack on hydrocarbon emissions.

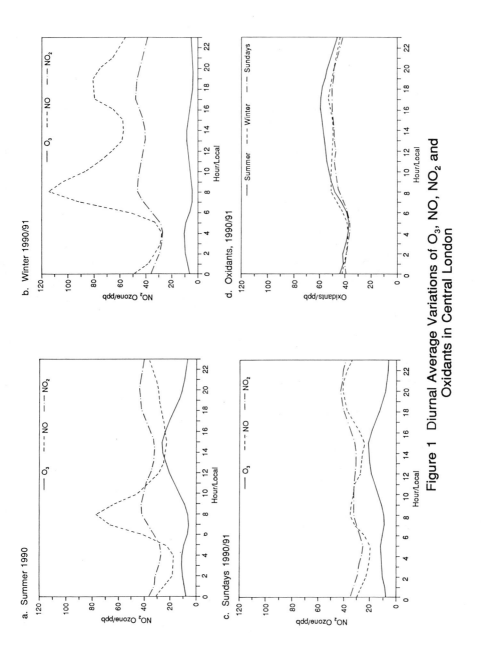

Figure 1 Diurnal Average Variations of O_3, NO, NO_2 and Oxidants in Central London

Although the NO_2 peak in summer appears to occur somewhat later than that for NO, this is unlikely - at this time of day and close to ground - to be strongly influenced by chemistry. The dominant influence on the timing of this peak is more likely to be vertical dispersion/exchange : NO_2 peaks after NO because NO, after emission, is being depleted by vertical mixing whilst NO_2 levels are increasing because of vertical mix-down from 'stale' air trapped aloft.

After the NO_2 peak, the balance between NO, NO_2 and O_3 shifts in favour of net ozone production, driven by the photolysis of NO_2. As a result, ozone concentrations reach their maximum in mid-afternoon. Further injection of NO during the early evening and the decline in insolation subsequently results in ozone levels falling whilst NO and NO_2 initially rise. This rise is less marked than during the corresponding morning rush-hour, due to the generally better-mixed conditions and hence improved dispersion, and increased ozone scavenging. As NO_x emissions decline, and NO and to some extent NO_2 are scavenged by ozone, levels of these species fall slowly through the night before the entire diurnal cycle recommences next day.

During the winter months (Fig 1b) overall NO_x levels are higher and O_3 concentrations lower. We again see a clear inverse NO/O_3 relationship, though the NO and NO_2 maxima are now coincident. This is because of the reduced vertical mixing in winter; furthermore the dominant NO_2 formation mechanisms in urban areas during winter - both fast - are ozone titration together with, possibly, catalytic surface oxidation at low temperatures (Lindquist et al, 1982). Increased NO_2 emissions are also possible from idling car engines in cold weather (Lenner et al, 1983).

Figure 1c shows that the diurnal variation of pollutant levels at Central London is characteristically different on Sundays, when traffic emissions are markedly reduced. The primary NO_x emission peaks are substantially lower; as a result, the corresponding ozone minima are absent and the resulting diurnal variation is very similar to that observed at rural ozone measurement sites.

In Figure 1d, Total Oxidant ($NO_2 + O_3$) concentrations at Central London are graphed for the periods examined in Figs 1a - 1c. It is clear that oxidant levels are always depressed at night, and decline steadily as ozone is dry deposited and both species are chemically scavenged. Primary NO_2 emissions and subsequent NO_2 formation from NO cause oxidant levels to rise rapidly during the morning - note, however, that this effect is much reduced at weekends. The photostationary state ensures that oxidant levels remain broadly constant during winter daytime hours. By contrast, during summer the oxidation of NO by peroxy radicals effectively bypasses this equilibrium state to produce excess NO_2 and hence O_3: this results in the rise in oxidant levels towards its observed mid-afternoon peak.

4   THE LONG-TERM O_3/NO_x RELATIONSHIP

In the previous section , the close relationship between short-term O_3 and NO_x concentrations in Central London has been clearly identified. The different O_3/NO_x balance prevailing in near-source urban areas of the UK may be readily demonstrated, over longer timescales, by examining recent measurements from the 17-site national ozone monitoring network managed by WSL. This network covers an intentionally broad range of UK location types from remote rural sites to the busy Central London location.

The differences in long term average O_3 concentrations at the various sites may be analysed in terms of the loss mechanisms operating at these locations. The reason for this is that, to a reasonable approximation, the *potential* for surface annual average O_3 concentrations due to production occurring in the troposphere and boundary layer (BL) is broadly similar over the whole of the UK. The resulting long-period surface O_3 concentrations at any point are then determined by loss mechanisms: the most important of these are reaction with NO (although it should be noted that this is not strictly a complete sink for O_3 during daylight hours) and dry deposition at the surface. As noted previously, the NO_x sink is likely to be the dominant loss process for O_3 in urban areas.

Continuous measurements of NO and NO_2 are available for many of the ozone network sites, including the urban and most high NO_x locations. Where unavailable, diffusion tube measurements of NO_2 from sites co-located with, or near to WSL primary network acid deposition sites (Cambell, 1988) have been used to derive long-term NO_x concentrations. To a reasonable approximation, total annual average NO_x levels at these sites may be taken to be 1.15 x NO_2 (assuming photostationary equilibrium, a reasonable premise at these rural and remote locations).

Measured and derived NO_x concentrations for UK measurement sites are graphed against corresponding average O_3 levels in Figure 2a. The period covers four full years of network measurements from 1987 to 1991, a substantially larger dataset than has been previously examined (Bower et al, 1989, i)); the number of sites examined is also significantly greater than in previous analyses.

The overall relationship in Fig 2a is clear. There is a strong ($r^2 = 0.88$) inverse relationship between long-term NO_x and O_3 levels across the UK, with the low O_3 concentrations at urban sites such as Central London corresponding to elevated NO_x levels, primarily arising from local motor vehicle emissions. The relationship for rural sites, at the low NO_x end of the graph, is less clearcut: this is due to the greater relative importance of the other major ozone sink mechanism - dry deposition - at these locations. It is interesting to note that the graph intercept of 28 ppb provides a reasonable geographically and seasonally averaged estimate of background ozone concentrations for the UK.

To examine the influence of dry deposition effects on long-term ozone levels, we graph in Fig 2b the relationship between annual average O_3 concentrations and C_D, the neutral geostrophic drag coefficient (Bower et al, 1989 i)). Values of C_D used here have been derived from a previous UK-wide analysis of surface roughness (Smith and Carson, 1977).

Figure 2d shows a firm relationship ($r^2 = 0.47$) between C_D and O_3 for rural sites only. Ozone concentrations at suburban and urban sites clearly lie well away from this best fit line, consistent with the NO_x sink being the dominant determinant of long-period O_3 at these sites. The scatter in the points for the rural sites reflects not only differences in the importance of the NO_x sink but also the surface component of the dry deposition process which will, of course, vary from site to site.

5   SHORT-TERM NO_2/NO_x RATIOS

Total NO_x levels at any location are dependent primarily on emissions and meteorology. Although the most important urban NO_x source - motor traffic - is

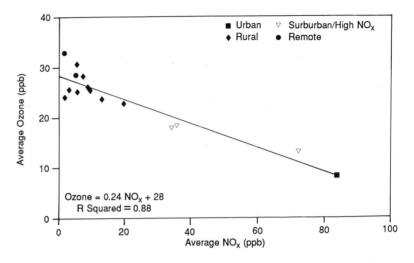

Figure 2a Scatter Diagram of Average UK Ozone vs Average NO$_x$, 1987-1991

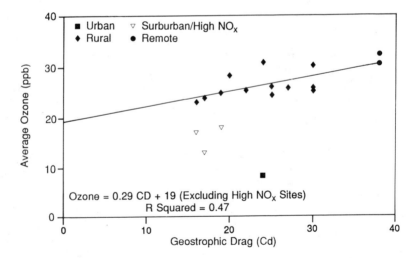

Figure 2b Scatter Diagram of Average UK Ozone vs Geostrophic Drag Coefficient, 1987-1991

reasonably constant throughout the year, other emissions from space heating and power stations rise during the winter months. Winter conditions are also associated with increased atmospheric stability, reduced mixing depths and consequently poorer dispersion of air pollutants. Taken together, these factors result in higher urban NO_x levels during the winter.

Atmospheric chemistry and source/receptor distance subsequently determine how NO_x is partitioned between NO and NO_2. A convenient measure of the extent and completeness of NO $\rightarrow$ NO_2 oxidation processes is provided by the measured NO_2/NO_x ratio by volume (R). Figures 3a - c graph the frequency distribution of hourly R values for Central London, Cromwell Road and Lullington Heath (a rural location in East Sussex). These figures demonstrate clearly how NO_2/NO_x ratios vary throughout the year and from site to site. Corresponding average R values are summarised in Table 1.

Throughout the year, measured R values at Central London consistently exceed corresponding NO_2/NO_x ratios for emissions (Fig 3a). These emission ratios are typically in the range 0.05 - 0.1, although somewhat higher emission ratios up to 0.3 are possible from warm idling car engines in cold weather conditions (Lenner et al, 1983). By comparison, measured NO_2/NO_x ratios at this site average 0.47 in winter and 0.59 in summer (Table 1). The high degree of NO oxidation typically observed at this background urban site is consistent with its relative distancing from emission sources. Enhanced NO $\rightarrow$ NO_2 oxidation due to increased ozone levels, together with higher temperatures and insolation rates, result in the higher average summer R values seen here. However, it should be noted that higher photolysis rates during the summer act against these processes and tend to result in a photostationary equilibriated state between NO, NO_2 and O_3.

Measured NO_2/NO_x ratios are markedly lower throughout the year at the kerbside Cromwell Road measurement site (Fig 3b). The distribution of R values during winter (average 0.17) is broadly consistent with that of primary emissions; for NO_2 concentrations exceeding 50 ppb (average R of 0.12), Fig 4a - a scatter diagram of hourly NO_2 versus NO_x concentrations during this season - demonstrates that the fit is even better. By contrast, during the summer months, R values are consistently higher, demonstrating substantial oxidation to have occurred. For instance, Fig 4b shows a substantial proportion of NO_2 concentrations at Cromwell Road during the summer months are associated with R values in the range 0.3 - 0.5.

It is somewhat surprising that a substantial summer/winter difference in measured NO_2/NO_x ratios can occur at such a kerbside location. This observation is, however, potentially explainable in a number of ways:

1)   Very fast oxidation processes are occurring in summer. Although catalytic surface oxidation of NO may occur very rapidly, both low temperatures and high NO concentrations are required for this mechanism to be effective. The reaction of NO with O_3 is, therefore, the likeliest fast mechanism for generating high NO_2 levels in kerbside environments in summer.

2)   Increased NO_2/NO_x emission ratios. Measurements, however, show such an enhancement of primary NO_2 levels is favoured in cold weather.

TABLE 1. - Summary of NO_2 to NO_x Ratios for Selected Sites

Site	Period	Summer			Winter		
		mean	STD	gmean	Mean	STD	gmean
Cromwell Road	All days	.24	.12	.21	.17	.12	.14
Central London	All days	.59	.16	.56	.49	.17	.46
Lullington Heath	All days	.85	.16	.84	.88	.14	.86
$NO_2 >= 50$ ppb							
Cromwell Road	All days	.23	.1	.21	.12	.06	.11
Central London	All days	.59	.18	.56	.36	.14	.34
Lullington Heath	All days	-	-	-	.82	.06	.82

TABLE 2. - Mean NO_2, NO_x, NO_2/NO_x and O_3 concentrations at selected sites during periods with high NO_2/NO_x ratios at Cromwell Road

Mean values		Summer 1990			Winter 1990/91	
Site	Pollutant	All data	CRD NO_2/NO_x > 0.36	CRD NO_2/NO_x > 0.36 and CRD $NO_2 >=$ 100 ppb	All data	CRD NO_2/NO_x > 0.29
Cromwell Road	NO_2	49.0	40.0	120.0	32.0	25.0
	NO_x	250.0	90.0	295.0	289.0	67.0
	NO_2/NO_x	0.24	0.47	0.4	0.17	0.4
	CO	2.9	1.2	4.8	3.4	0.9
Central London	NO_2	36.0	32.0	56.0	40.0	23.0
	NO_x	69.0	44.0	71.0	104.0	35.0
	NO_2/NO_x	0.59	0.74	0.8	0.49	0.68
	CO	0.9	0.6	1.2	1.4	0.6
	O_3	13.0	20.0	64.0	15.0	7.0
Lullington Heath	O_3	37.0	39.0	80.0	21.0	25.0

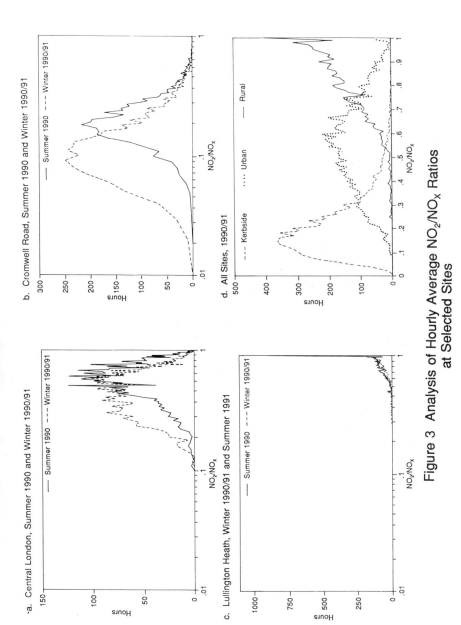

Figure 3 Analysis of Hourly Average NO$_2$/NO$_x$ Ratios at Selected Sites

a. Winter 1990/91

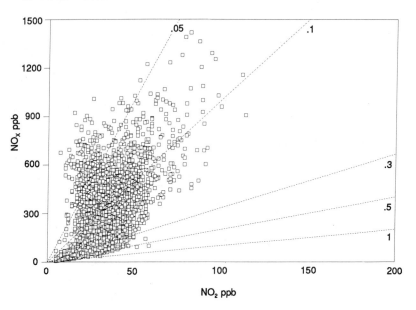

b. Summer 1990

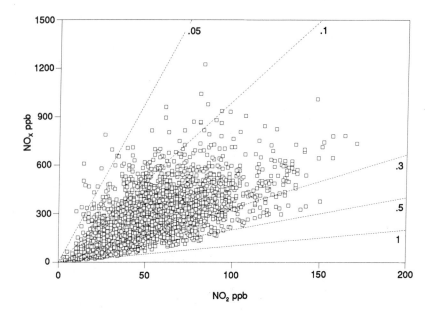

Figure 4 Scatter Diagrams of Hourly Average NO$_2$ and NO$_x$
Concentrations at Cromwell Road

3)   Increased relative NO_2 levels, reduced NO_x and higher resulting NO_2/NO_x ratios resulting from 'aged' airmasses advected from elsewhere.

To attempt to further investigate the cause of enhanced NO_2/NO_x ratios at Cromwell Road, we summarise in Table 2 the average CO, NO_2, NO_x and NO_2/NO_x ratios for complete seasonal periods at this site, together with corresponding statistics for periods of above - average R (R exceeding its mean value by >1σ). Pollutant concentrations at Central London are also tabulated, together with ozone levels here and at a rural site in S E England.   These statistics are also summarised for periods of elevated R ratios at Cromwell Road combined with elevated (> 100 ppb) NO_2 concentrations at this site. This analysis is only possible for the summer season, since there are very few occasions in winter when both conditions apply.

Table 2 suggests that the high R values at Cromwell Road are for the most part associated not with enhanced NO_2 but with depressed NO_x (and CO). This observation is not explainable by increased NO → NO_2 oxidation or elevated emission ratios (mechanisms 1 and 2 above), but is more consistent with 'aged' - and therefore well oxidised - urban air.

Somewhat different conditions are associated with occasions when both NO_2/NO_x ratios and absolute NO_2 concentrations are high at Cromwell Road. Table 2 shows this situation to occur when ozone concentrations are significantly elevated (average of 64 ppb at Central London, 80 ppb at Lullington Heath): this clearly demonstrates ozone titration to be the cause of elevated NO_2 concentrations and therefore increased NO_2/NO_x ratios).

Taken together, these observations show that elevated NO_2/NO_x ratios at a kerbside location during the summer are often associated with aged, well oxidised, urban air: however, when O_3 concentrations are elevated, the fast reaction of primary NO can result in considerable NO_2 formation.

In complete contrast to this analysis of a kerbside urban environment, Fig 3c graphs NO_2/NO_x ratios for a rural site in S E England. NO → NO_2 oxidation processes will exert a strong influence at remote, rural locations distanced from primary emissions. In this circumstance, resulting NO_2/NO_x ratios tend towards their daytime photostationary equilibrated value (for O_3 ~ 30 ppb) of ~ 0.85. As is readily apparent from Fig 3c, R values at Lullington Heath are substantially larger than those at the urban sites previously discussed; even during the winter months, R approaches or exceeds 0.85 for much of the time and rarely falls below 0.6. Clearly, therefore, photostationary equilibrium, or a near-equilibrated state between O_3 and NO_x appear to exist for most of the time at such rural locations.

Graphing the annual NO_2/NO_x distribution functions for Cromwell Road, Central London, and Lullington Heath on linear axes, the trend for these ratios to increase from kerbside to urban and rural locations is readily apparent (Fig 3d). It is clear that measured NO_2/NO_x distribution are here providing a good indication of source/receptor distances.

The distribution function for Cromwell Road is markedly skewed, and apparently lognormal. This skewed distribution in near-field situations is due to a lower limit to a measured NO_2/NO_x ratios being set by those typically prevailing in emissions (~ 0.05 - 0.1). The distribution function for the intermediate-field Central

London site is, by contrast, approximately normal. For the far-field rural Lullington Heath site, it should be noted that measured NO concentrations are below the analyser detection limit - normally 0.5 ppb - for roughly 25% of the time (mostly at night), leading to a large proportion of measured NO_2/NO_x ratios of unity.

6   ANNUAL AVERAGE NO_2/NO_x RATIOS IN THE UK

We have seen in the previous section that hourly average NO_2/NO_x ratios provide a good qualitative measure of the degree and completeness of primary $NO \rightarrow NO_2$ oxidation processes. As such, these ratios can vary substantially, both from site to site and throughout the year. In Figure 5, graphed annual average NO_2 and NO_x concentrations enable the relationship between these species to be assessed over longer time scales. Measured concentrations from 1976 are included for WSL - operated UK monitoring stations with annual data capture rates exceeding 50%. For clarity, sites plotted are classified as rural, suburban (sites such as Stevenage), urban (including Central London and UK NO_2 Directive Monitoring Stations) and kerbside (including Cromwell Road). As can be seen, data availability is limited for rural and remote sites, although continuous measurements commenced during 1990 at three such locations.

Interestingly, the site groups graphed in Fig 5 appear to cluster into well-defined NO_2/NO_x populations. For the highest UK NO_x levels, measured at kerbside sites such as Cromwell Road, the atmosphere's capacity to oxidise NO to NO_2 is limited, primarily by ambient ozone availability. The NO_2/NO_x relationship is highly nonlinear for such sites, and most NO_2 annual averages are in the range of ~ 40 - 50 ppb, apparently independent of corresponding long-term NO_x concentrations.

At the same time, data for urban and suburban sites typically display a broadly linear NO_2/NO_x relationship with annual R values of ~ 0.4. If we assume a photostationary state, this figure would be appropriate to an ozone concentration of 15 ppb, a reasonable expectation for long-term average concentration at these sites (Bower et al, 1991 (i)). This apparent agreement should, however, be regarded with some caution: although it is appropriate to assume equilibrated airmasses in rural areas, this will clearly not apply in urban areas with local emission sources.

It is, however, interesting to note that corresponding data for rural sites display annual R values of ~ 0.8, appropriate to a photostationary equilibrated state for typical UK background ozone concentrations of ~ 30 ppb. This is surprising since, even in a well-mixed airmass, NO_2 and NO_x concentrations will at best be controlled by the photostationary equilibrium during daytime hours only. The night-time $NO + O_3$ reaction dominates for half the time on average, and this will tend to result in long-term average NO_2 concentrations higher than those predicted by the equilibrium relationship. In practice, however, local primary NO_x emissions make some contribution even at rural sites, and these will tend to push data to the higher NO_x side of the equilibrium line. It appears, therefore, that these two mechanisms effectively balance out in urban and rural locations, to produce the apparent linear NO_2/NO_x relation seen in Figure 5.

7   TREND ANALYSIS OF NO_x, NO_2 AND O_3 CONCENTRATIONS

In previous sections the relationship between NO_x, NO_2 and O_3 concentrations in urban (and other) environments has been explored, without considering possible

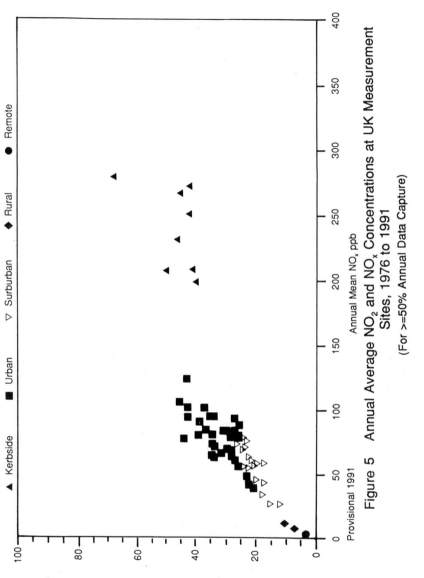

Figure 5   Annual Average NO₂ and NOₓ Concentrations at UK Measurement Sites, 1976 to 1991

(For >=50% Annual Data Capture)

long-term changes over time in these parameters. Only a few air monitoring sites in the UK have sufficiently extended datasets to allow meaningful trend analysis. Two of these are the urban Central London and the kerbside Cromwell Road stations considered in this paper. For instance, ozone measurements commenced at Central London in 1972, whilst NO_x measurements date from 1976. Long-term pollutant trends at these locations are now assessed, together with - for comparative purposes - NO_x and O_3 results from the suburban Stevenage location (some 30 miles to the north of London) and ozone data from the rural Sibton site (in Suffolk).

Air quality trends deduced from time series analysis should always be regarded with caution. Changes in monitoring methodologies, instrumentation and operational practice always occur over extended measurement periods. Also of significance is the fact that UK gas primary reference standards and calibration methodologies have evolved and advanced considerably over time, resulting in substantial improvements in measurement accuracy, precision and reproducibility.

Some of the measurements analysed here date back to the commencement of automatic air monitoring in the UK and incomplete data capture rates during some years should be recognised as a possible 'masking' factor when attempting to derive long-term trends.

Notwithstanding these issues, we summarise in Table 3 best-fit regression line slopes, slope standard errors and correlation coefficient derived from NO_x, NO_2 and O_3 time series for the sites under analysis. To elucidate possible trends in average and peak concentrations, both annual data and 98th percentile hourly average concentrations for each year are considered. Time series for one of the datasets (Central London, 98th percentile) are graphed in Figure 6.

It is clear from Table 3 that there is a considerable scatter for some datasets, much of this resulting from year to year variations in concentrations. Consequently, there is often a high degree of uncertainty in the calculated regression line slopes. For the purposes of subsequent analyses, we identify pollutant trends as significant when the magnitude of the derived regression line slope exceeds the standard error in that slope estimate. Significant trend directions for both average and peak concentrations are summarised in Table 4.

All the sites here considered are in the SE of England. This is an area of increasing NO_x emissions, primarily due to a rise in road traffic. Whilst power station emissions remained broadly constant over the period from 1970 to 1989, those from road transport have increased steadily from 0.7 to 1.3 M Tonnes Yr^{-1} (Leech, 1991). In this context, it is somewhat surprising to note that the calculated NO_x regression line slopes for the kerbside Cromwell Road site, although positive for both average and peak measured concentrations ($\sim +2$ ppb yr^{-1}), are associated with particularly large data scatter and cannot therefore be regarded as statistically significant (Tables 3 and 4). At the same time, however, significant overall increases in NO_x concentrations are apparent for the Central London (98th percentile only) and Stevenage sites. The upward trend is particularly marked at Stevenage, (with increases of 2 and 12 ppb yr^{-1} for average and peak levels), and this is probably due to increasing traffic densities on the nearby A1M Motorway.

Previous analyses have demonstrated the NO_x sink to be the dominant loss mechanisms for O_3 in urban and many rural areas. As a result, we would expect

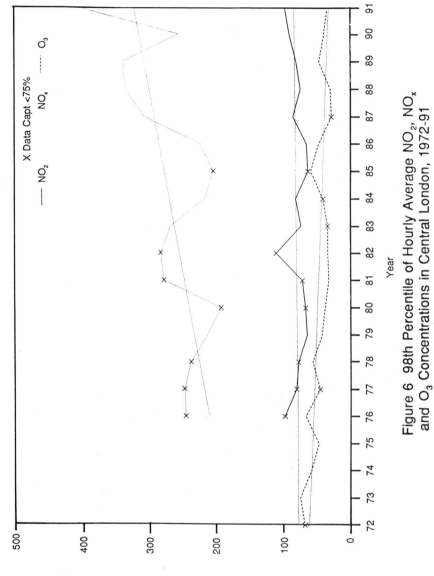

Figure 6 98th Percentile of Hourly Average NO_2, NO_x and O_3 Concentrations in Central London, 1972-91

TABLE 3. - NO_2, NO_x and O_3 Trends at Representative Locations

	Grad ppb/year	Std err ppb/year	Grad %/year	Std err %/year	Corr. Coeff.
CRD mean NO_2	-1.01	.83	-2.06	1.70	.34
CRD mean NO_x	2.01	2.90	.88	1.27	.20
CRD 98% NO_2	-3.59	2.87	-2.85	2.28	.35
CRD 98% NO_x	1.79	10.10	.26	1.45	.05
CLL mean NO_2	-.29	.25	-.75	.65	.29
CLL mean NO_x	-.53	.72	-.63	.85	.19
CLL mean O_3	-.25	.11	-2.25	.95	.48
CLL 98% NO_2	.34	.75	.43	.94	.12
CLL 98% NO_x	7.54	2.64	2.84	1.00	.61
CLL 98% O_3	-1.52	.40	-2.33	.61	.67
STE mean NO_2	.29	.15	1.25	.64	.48
STE mean NO_x	1.70	.39	2.77	.63	.77
STE mean O_3	-.28	.25	-1.93	1.72	.29
STE 98% NO_2	.23	.47	.42	.85	.14
STE 98% NO_x	12.00	2.32	4.51	.87	.82
STE 98% O_3	-.04	.84	-.08	1.66	.01
SIB mean O_3	-.47	.19	-1.85	.75	.52
SIB 98% O_3	-1.41	.57	-2.17	.87	.53

TABLE 4. - Summary of NO_2, NO_x and O_3 Trends

	NO_2		NO_x		O_3	
	AV.	98% ile	AV.	98% ile	AV.	98% ile
CRD	↓	↓	-	-	N.A	N.A
CLL	↓	-	-	↑	↓	↓
STE	↑	-	↑	↑	↓	-
SIB	N.A	N.A	N.A	N.A	↓	↓

(Trend shown if magnitude of slope > standard error)

long-term ozone trends at these sites to be determined primarily by corresponding NO_x variations. Tables 3 and 4 confirm this, and demonstrate significant downward ozone trends at all locations, with overall annual average concentrations falling by ~ 2% per annum. It should, however, be recognised that ozone trends in other parts of the UK, outside SE England, will not be as strongly or directly influenced by changing NO_x emissions from motor vehicles. At remote (far-field) locations, O_3 trends will be influenced more by total UK NO_x emissions as well as factors such as meteorology and topographic/locational influences. Ozone concentrations at these locations may not, therefore, exhibit the downward trend demonstrated for the sites considered here.

Clearcut NO_2 trends are not readily apparent. An upward trend in annual averages is apparent at Stevenage (+ 1.25% yr^{-1}) whilst a decline is evidenced at Cromwell Road (- 2% yr^{-1}) and Central London (- 0.75 % yr^{-1}). These observations are not readily explainable at the present time.

8   A RECENT POLLUTION EPISODE IN LONDON

A major winter pollution episode occurred in London during the second week of December 1991, with unusually high levels of NO_2 being observed from the 12th to 15th. Details of the episode are here published for the first time.

UK-wide air quality measurements show the episode to have been confined to London and surrounding areas. Meteorological conditions throughout the period were cold and foggy, with low wind speeds and stable, anticyclonic synoptic conditions. These conditions are typical of winter stagnation - type air pollution episodes, and are similar to those encountered during major smoke/SO_2 smogs of the 1950's and 1960's. It is interesting to note, in passing, that these also tended to occur in the first half of December.

Figure 7 graphs measured levels of NO_x, NO_2, CO and SO_2 at Central London during the episode period. Observed pollutant concentrations were very similar at Cromwell Road and a monitoring station in West London. Peak NO_2 concentrations at Central London reached a maximum hourly average of 423 ppb at 0600 hours on the 13th - this was the highest NO_2 level recorded in UK urban (non-kerbside) environments since measurements commenced in 1976. Although corresponding NO_2 concentrations at Cromwell Road simultaneously peaked at 382 ppb, it may be noted that higher levels have been observed previously at this location (1817 ppb in 1982, 591 ppb in 1976).

Throughout the episode, measured NO_2 and NO_x concentrations were closely coupled at all London measurement sites. O_3 concentrations were extremely low, with any background concentrations being overwhelmed by primary NO_x emissions. As a result, NO_2 levels were directly controlled by emissions, with low NO_2/NO_x ratios in the range 0.1 - 0.3 at Central London and ~ 0.07 - 0.2 at Cromwell Road. These values are broadly consistent with expected motor vehicle emission ratios for slow/congested traffic during cold weather conditions.

It may be noted that CO levels were also elevated and, not surprisingly, these tracked very closely with measured NO_x (Fig 7). The observed NO_x/CO ratios during this period are, again, consistent with vehicle emissions (~ 10 : 1 by mass at 20 Kph). However, SO_2 levels were low and did not correlate well with those of

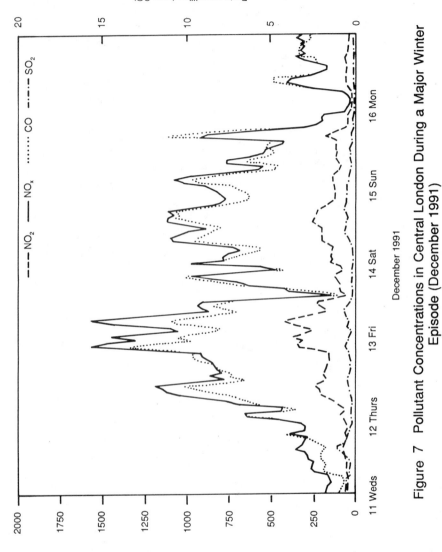

Figure 7   Pollutant Concentrations in Central London During a Major Winter
Episode (December 1991)

CO or NO_x. These observations clearly identify motor vehicle emissions as the dominant NO_x source during the episode.

This episode is of considerable interest for a number of reasons, not least because of the extremely high levels of pollution recorded. Its sharp cut off in both space and time, together with the fact that levels of only one major respiratory irritant (NO_2 but not SO_2) were elevated, also make the episode potentially interesting in terms of assessing possible health implications for the population at large.

9 SUMMARY AND CONCLUSIONS

1.   Ambient measurements of NO_2, NO_x and O_3 are reviewed at two long-running London monitoring sites operated by Warren Spring Laboratory for Department of the Environment. These are a background urban site in Victoria, Central London and a kerbside location in Cromwell Road.

2.   Levels of these pollutants are strongly interrelated in urban areas. The photostationary state between NO, NO_2 and O_3 is, however, significantly perturbed by local NO_x emissions, insufficient transport times to allow equilibrated conditions, and a variety of NO_2 formation mechanisms not involving direct O_3 loss. Furthermore this state can, at best, only be applicable during daytime hours.

3.   The NO_x sink is the dominant O_3 loss process wherever local NO concentrations are comparable to the background O_3 boundary layer concentration of ~ 30 ppb. This is the case for most urban areas of the UK.

4.   The strong interaction between NO_2, NO_x and O_3 is readily apparent when examining diurnal variations of these pollutants in Central London. These are strongly influenced by the NO_x emission cycle for motor traffic, which peaks at rush-hour periods. Characteristically different diurnal patterns in measured pollutant concentrations are observed in different seasons of the year, due primarily to changing meteorological and dispersion conditions. The effects of lower primary NO_x emission levels on Sundays are also clear from these analyses.

5.   There is a strong inverse relationship between measured annual average O_3 and NO_x concentrations. This applies at a wide range of UK measurement sites, including urban, suburban and rural locations. The relationship is less clearcut at rural sites, due to the greater relative importance of dry deposition as an ozone sink in such areas.

6.   Atmospheric chemistry and source/receptor distances play an important role in determining how NO_x in urban areas is partitioned between NO and NO_2. Analyses of the NO_2/NO_x ratio (R) for measured hourly average concentrations show these to be significantly higher in summer than in winter, and lower in near-source areas than in the far-field.

7.   Winter NO_2/NO_x ratios at a kerbside site are here shown to be broadly consistent with vehicle emission ratios. It is, however, interesting to observe a significant summer/winter difference in measured R values. Further analyses demonstrate that NO_2/NO_x ratios exceeding the likely

range in primary emissions can be produced, when ozone levels are elevated, by the fast NO/O_3 reaction.

8.   An analysis of annual average NO_2/NO_x ratios at all UK measurement sites show these cluster into well-defined populations according to location type. Annual average NO_2 levels at kerbside sites are mostly in the range 40 - 50 ppb, apparently independent of corresponding NO_x concentrations. This is due to the atmosphere's limited capacity for oxidising large amounts of NO to NO_2 at such locations. A linear relationship between annual average NO_2 and NO_x is, however, apparent for most urban and suburban locations. Annual average R values of ~ 0.4 are observed at these sites, appropriate to a photostationary state for long-term ozone concentrations of ~ 15 ppb. Similarly, observed R values of ~ 0.8 for rural locations are consistent with annual average O_3 levels ~ 30 ppb.

9)   Trend analyses of measured annual average pollutant concentrations have been undertaken for long-running kerbside, urban, suburban and rural monitoring sites. Increasing NO_x levels are evidenced at Central London and the suburban Stevenage sites, although no clear trend is apparent at Cromwell Road. The downward trend in long-term ozone concentrations, observed at all locations studied in SE England, is consistent with increasing emissions and rising ambient concentrations of NO_x.

10) A recent high NO_x episode in London has been analysed. The highest urban-non-kerbside NO_2 concentration ever observed in the UK (423 ppb in Central London) was recorded in December 1991. Both NO_2 and NO_x levels were strongly elevated, and closely coupled, throughout London during the period from December 12 to 15. CO levels were also high, although O_3 and SO_2 were markedly depressed. The episode occurred during a period of unusually cold stable anticyclonic weather conditions and analyses demonstrate the major pollutant source to have been motor vehicle emissions.

10   ACKNOWLEDGEMENT

The UK national urban and rural air monitoring networks are funded by the Department of the Environment and are part of their ongoing research programme, (the UK Air Pollution Core Programme PECD 7-12-29, incorporating PECD 7-12-28).

11   REFERENCES

1.   Leech, P.K. UK Emissions of Air Pollutants 1970 - 89. Warren Spring Laboratory Report LR 826 (AP), 1991.

2.   Simpson, D. An Analysis of Nitrogen Dioxide Episodes in London. Warren Spring Laboratory Report LR 575 (AP), 1987.

3.   Simpson, D., Perrin, D.A., Varey, J.E., and Williams, M.L. Dispersion Modelling of Nitrogen Oxides in the United Kingdom. Warren Spring Laboratory Report LR 693 (AP), 1988.

4. Lampert, J.E. The Automatic UK Air Quality Monitoring Networks: Infrastructure and Site Locations. Warren Spring Laboratory Report LR 813 (AP), 1991.

5. Bower, J.S., Stevenson, K.J., and Broughton, G.F. et al. Ozone in the UK: A review of 1989/90 Data. Warren Spring Laboratory Report LR 793, 1990.

6. Bower, J.S., Broughton, G.F. and Dando, M.T. et al. Urban NO_2 Concentrations in the UK in 1987. Atmos. Environ., 25B, 2, 267-283, 1991.

7. Bower, J.S., Lampert, J.E., and Stevenson, K.J. et al. A Diffusion Tube Survey of NO_2 Concentrations in Urban Areas of the UK. Atmos. Environ., 25B, 2, 255-265, 1991.

8. Bower, J.S., Broughton, G.F., and Dando, M.T. et al. Surface Ozone Concentrations in the UK in 1987-88. Atmos. Environ., 23, 9, 2003-2016, 1989.

9. Williams, M.L., Broughton, G.F., and Bower, J.S. et al. Ambient NO_x Concentrations in the UK: 1976-84 - A Summary. Atmos. Environ., 22, 12, 2819-2840, 1988.

10. Broughton, G.F. and Bower, J.S. et al. Air Quality in the UK: A Summary of Data for 1990/91. Warren Spring Laboratory Report LR 883, 1992.

11. Sweeney, B.P. and Stacey, B. Intercomparison and Intercalibration Techniques employed for the UK National Air Monitoring Networks. Warren Spring Laboratory Report LR 812 (AP), 1991.

12. Lindquist, O., Ljungström, E., Svenson, R. Low Temperature Thermal Oxidation of Nitric Oxide in Polluted Air. Atmos. Environ., 16, 8, 1957-1972, 1982.

13. Lenner, M., Lindquist, O., Rosen, A. The NO_2/NO_x Ratio in Emissions from Gasoline-Powered Cars : high NO_2 percentage in Idle Engine Measurements. Atmos. Environ., 17, 8, 1395 - 1398, 1983.

14. Cambell, G.W. Measurement of Nitrogen Dioxide Concentration at Rural Sites in the United Kingdom using Diffusion Tubes. Environ. Poll., 55, 251 - 270, 1988.

15. Smith, F.B. and Carson, D.J. Some Thoughts on the Specification of the Boundary Layer Relevant to Numerical Modelling Boundary Layer Met., 12, 307-330, 1977.

16. Derwent, R.G. and Nodop, K. Long Range Transport and Deposition of Acidic Species in North West Europe, Nature, 324, 356-358, 1986.

Measurements of Rural Photochemical Oxidants

G. J. Dollard and T. J. Davies

AEA TECHNOLOGY, HARWELL LABORATORY, DIDCOT, OXFORDSHIRE OX I I ORA, UK

1 INTRODUCTION

Interest in photochemical oxidants in the troposphere goes back many years principally with concern for the environmental impact of photochemically derived ozone. Measurements and studies relating to the Los Angeles smogs during the 1950s indicated that the formation of ozone involved reactions between pollutants that were driven by the energy of the sun. Since these early studies, continuing work has shown that ozone formation occurs on a regional scale in many areas of the globe. Ozone can affect adversely vegetation, animal and human health[1,2]. In addition to ozone other oxidising species may be formed through photochemical pathways in polluted air. Two key species are hydrogen peroxide (H_2O_2) and peroxyacetyl nitrate (PAN).

Hydrogen peroxide is an important trace oxidising gas in the atmosphere; it is very soluble ($K_H \sim 10^5$ mol l^{-1} atm^{-1}) and thus freely enters solution in cloud and rain water. The potential for hydrogen peroxide to act as a major oxidant for sulphur dioxide has been recognised for a number of years and has been the subject of much experimental work[3,4,5].

PAN is formed uniquely by photochemical reactions in polluted air in which it may act as a reservoir facilitating the long range transport of nitrogen. PAN is not very soluble in water and it is not removed efficiently from the atmosphere by rainfall. Both PAN and H_2O_2 are strong oxidising agents that may be dry deposited at the earth's surface where they may cause damage to vegetation.

The formation of O_3, PAN and H_2O_2 in the atmosphere is summarised schematically in figure 1. Ozone formation consists of the recombination of atomic and molecular oxygen (reaction 2); at lower (tropospheric) levels in the atmosphere the main source of atomic oxygen is the photolysis of nitrogen dioxide (1). As can be seen the ozone generated by this mechanism may react with nitric oxide to reform NO_2 (3). During hours with sunlight a photochemical steady state is achieved and the ozone concentration is given by expression (4). It is evident that emissions of nitric oxide at the surface may destroy the ozone produced through reaction (3). Additionally, at night time, with the ozone formation switched off, the ozone concentration may fall due to reactions with NO and through dry deposition at the surface.

In an atmosphere containing hydrocarbons, primarily from pollutant emissions or in some cases natural emissions from vegetation, nitric oxide may be converted to NO_2 by reaction (5), thereby facilitating net production of ozone. From figure 1 it may be seen that it is the peroxyradical (RO_2) that initiates this reaction sequence; this radical is generated during the degradation of hydrocarbons following attack by hydroxyl (OH) free radicals. RO_2 may also react directly with NO_2 to form organic nitrogen compounds - peroxy nitrates (9). In the case when the R fraction is the acetyl group, peroxyacetyl nitrate is formed which, generally speaking, because of its thermal stability is the most important of this type of compound generated in the atmosphere.

The formation of another free radical species the hydroperoxy radical (HO_2) is also shown in figure 1. The self reaction of HO_2 (10) as shown is the most significant route for the formation of hydrogen peroxide. The scheme depicted in figure 1 illustrates the common origins of the three oxidant species.

It is clear that the measured concentrations of photo oxidants in the ambient atmosphere will be influenced by the rates at which the chemical mechanisms proceed. In addition to these chemical factors, large and local scale mixing processes that occur in the troposphere can exert strong influences on ozone concentrations measured at ground level.

During the daytime the surface boundary layer has a depth of approximately 1 km, sustained by convective and turbulent mixing processes, driven by solar heating and drag at the earth's surface. Within this daytime boundary layer any pollutants or photo oxidants will be well mixed. With the onset

Figure 1.   Simplified scheme showing the formation of O_3, PAN and H_2O_2 in the troposphere.

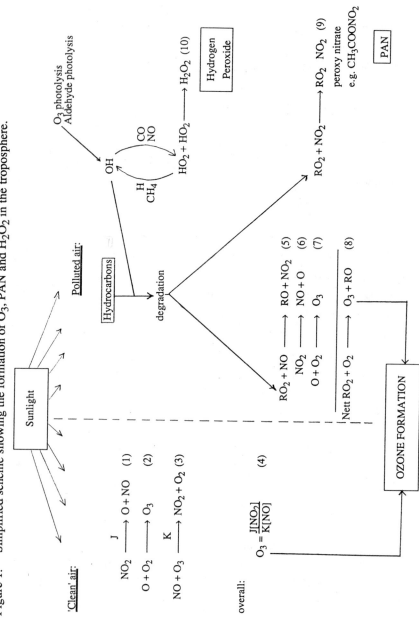

of sunset and subsequent cooling at the surface, the depth of the boundary layer may contract to perhaps a few tens of metres above the surface. Within the shallower boundary layer pollutants trapped beneath the capping temperature inversion become isolated and dry deposition to the surface depletes the concentration near to the surface; hence night time concentrations are observed to decline. In the particular case of ozone, emissions of NO at the surface can destroy ozone through reaction (3). With the onset of sunrise convective processes break down the nocturnal stratification; following this, pollutants are mixed down to the surface and their concentrations may be seen to rise. If an extended series of ground based ozone concentration measurements are plotted, certain characteristic features may be discerned. Figure 2 illustrates one such plot, this annotated figure represents ozone data recorded at the Devilla site in Central Scotland[6]; the three lines illustrate the relative contributions, at ground level, of the main ozone sources.

The stratosphere is an important source of ozone[1], and from figure 2 it may be seen that it influences in two ways ozone observations recorded at ground level. Firstly, it contributes to the general background level within the troposphere; this ozone is transported principally into the troposphere through exchange processes occurring at the equator and at the poles, as well as slow diffusion across the tropopause. Secondly, occasional spikes in the ozone concentration are seen, these are due to penetration of stratospheric air directly to the surface.

Superimposed upon the 'background' concentration is the tropospheric ozone generated by the mechanisms discussed above and summarised in figure 1. Within this component are spikes representing short periods, (upto a few days) when ozone concentrations may be very high. These episodes occur when conditions are particularly favourable for ozone formation. The conditions favourable to such enhanced production of ozone (and other photo oxidants) are those that optimise conditions and allow photochemical processes to proceed apace. These include factors that influence the presence of precursor compounds, sunlight and the physical processes that control transport and dispersion.

Typically for the UK such conditions tend to occur during anticyclonic weather conditions during the spring and summertime. These conditions produce warm temperatures, zero or thin cloud cover, light winds and easterly air flow. These light easterly winds transport pollutant precursors (NO_x and hydrocarbons) from mainland Europe to the UK. Such conditions may persist for several days

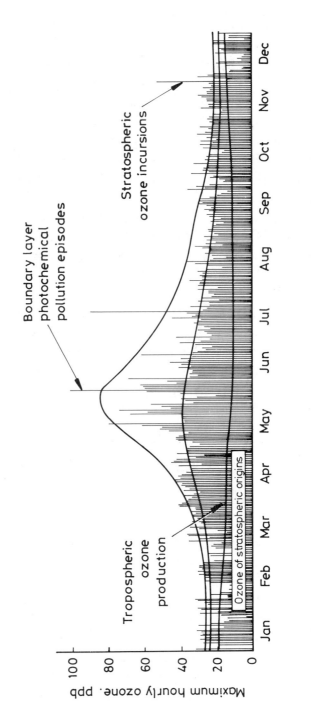

Figure 2. Contribution from the major ozone sources to the daily maximum hourly mean at the ground.

owing to the stability of the high pressure system. During this time photo oxidant concentrations may rise very markedly. Usually the breakdown of the weather system leads to a flushing out of the boundary layer and concentrations return to the 'normal' levels.

Thus the patterns of concentration observed in surface data on photo oxidant species are generated by interactions between chemical and physical processes occurring in the atmosphere. In addition, topographical features may influence observed levels. Thus sites on elevated ground may be exposed more frequently to free tropospheric air as the site may at times be above the boundary layer or less frequently capped by nocturnal stability. Similarly land sea breezes may serve to sustain mixing processes at coastal sites. In both these cases there may be a dampening of the diurnal cycle leading to increased daily average concentrations. To summarise, ozone, PAN and H_2O_2 data from any one site exhibit characteristics generated by chemical, topographical and meteorological factors.

UK Photo Oxidant Data

All of the ozone data summarised in this paper was collected from numerous research laboratories and Universities as part of the activity of the UK Photochemical Oxidant Review Group (PORG). The ozone data has been compiled into a database at AEA Technology Harwell from where it is available on floppy disc. Further information on how to obtain copies of the database may be obtained from the authors. Figure 3 summarises the site locations, for which further details are given in the interim PORG ozone report[2]. At present there are about 47 sites held in the database with a total of 237 site years. Of the 47 sites, 17 are linked by telemetry and constitute the National Rural Network which was set up and funded by the Department of Environment following recommendations by PORG. At all of the PORG sites commercial instruments are used to monitor continuously ozone concentrations. The most widely used method utilises a photometric instrument which measures the concentration of ozone by monitoring the absorption of ultra-violet light. In addition, instruments utilising the detection of chemiluminescence generated when ozone reacts with ethylene have been used. Full details of methods used are described in the interim PORG report[2].

In comparison to ozone there are only relatively few sites where measurements of PAN and H_2O_2 have been made. These are illustrated in

No.	Site Name	Grid Ref.
*1	Strath Vaich	NH347750
3	Devilla	NS957894
4	Glasgow	NS554678
*5	Bush	NT245635
*7	Eskdalemuir	NT235028
*8	Great Dun Fell	NY711322
*9	Wharleycroft	NY698247
10	Wray	SD619678
12	Stodday	SD462587
13	Hazelrigg	SD492579
*14	High Muffels	SE776939
15	Yorkminster	SE603522
*17	Glazebury	SJ690959
*18	Ladybower	SK164892
19	Jenny Hurn	SK817986
20	West Burton	SK804864
21	Brampton	SK843810
22	Thorney	SK858731
23	Lincoln	SK983729
24	Syda House	SK312696
*26	Bottesford	SK797376
*27	Aston Hill	SO298901
*28	Sibton	TM364719
*29	Stevenage	TL237225
30	St. Osyth	TM104183
32	Chigwell	TQ442919
33	Hainault	TQ460917
*36	Harwell	SU474863
37	Canvey Island	TQ782847
38	Islington	TQ321831
39	St. Bartholomew's	TQ319821
40	Natwest Tower	TQ331814
41	County Hall	TQ306797
*42	Central London Lab.	TQ291790
43	Cromwell Road	TQ264789
44	Kew	TQ185779
47	Teddington	TQ156706
49	Ascot	SU946688
54	East Malling	TQ712572
*59	Lullington Heath	TQ538016
*60	Yarner Wood	SX786789
*62	Lough Navar	IH065545
64	Mace Head	IL740320
65	Clatteringshaws	NX553779
66	Dursley	ST755967
67	Fawley	SU474202
68	West Beckenham	TG141388

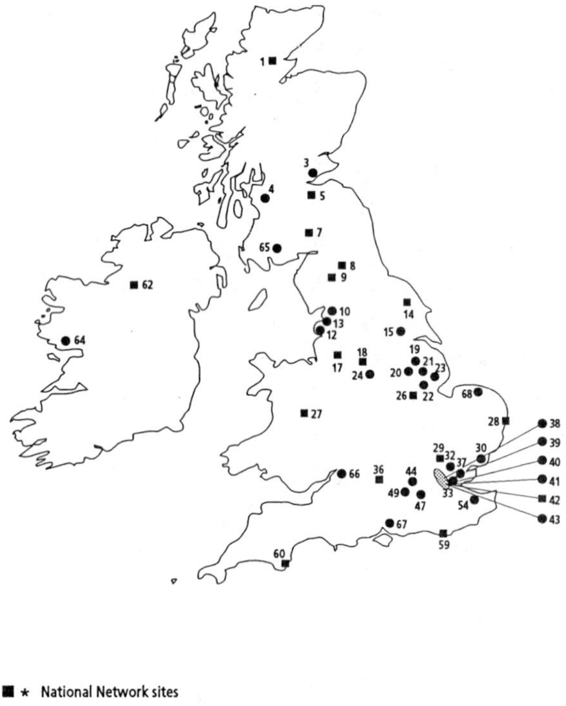

■ ★ National Network sites

Figure 3.   Ozone measurement sites within the UK PORG database.

figure 4; for the purpose of the present report only the UK data collected at AEA Technology Harwell will be presented.

The measurements of hydrogen peroxide at Harwell were made using a chemiluminescence technique, the details of this method are described elsewhere[7]. Briefly, H_2O_2 is stripped from the air in a mixing coil and determined from the chemiluminescence generated when H_2O_2 reacts with luminol in the presence of a microperoxidase catalyst.

The equipment and detection system for the routine measurement of PAN is described in full by Dollard et al.[7] Briefly, an automatic electron capture gas chromatograph was utilised. A 6 cm^{-3} sample of ambient air was injected onto a 45 cm long 0.6 cm i.d. glass column packed with 5% polyethylene glycol (PEG400) on diatomite (100-200 mesh).

Both the PAN and the H_2O_2 analysers at Harwell were housed in a building on a former airfield away from other laboratory buildings and close to the National Rural Network ozone monitoring site depicted in figure 3.

2   RESULTS AND DISCUSSION

Because of the large amount of data collected on these three species only general features of the data will be discussed. For more detailed information on ozone the interim PORG report[2] is a most useful document at present; a follow up report on ozone is currently in preparation and is expected to be published by the end of 1992. The scheduled PORG report will also contain information and discussion on PAN and H_2O_2. More details of the PAN and H_2O_2 measurements made at Harwell are available in several published reports[7,8,9].

From the preceding discussions on the chemical and physical factors contributing towards the observed photo oxidant concentrations at ground level it is apparent that daily, seasonal and possibly long term trends in data may be identified.

Examination of data for all three species indicate that strong diurnal variations in concentration occur at many sites. Figure 5 illustrates hourly ozone data averaged over the year for the Harwell site for 1988 and 1990. The clear

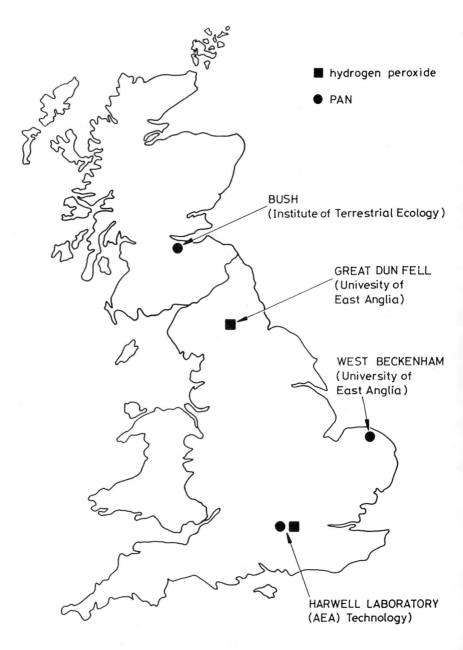

Figure 4. Locations of measurement sites for H$_2$O$_2$
and PAN in the UK.

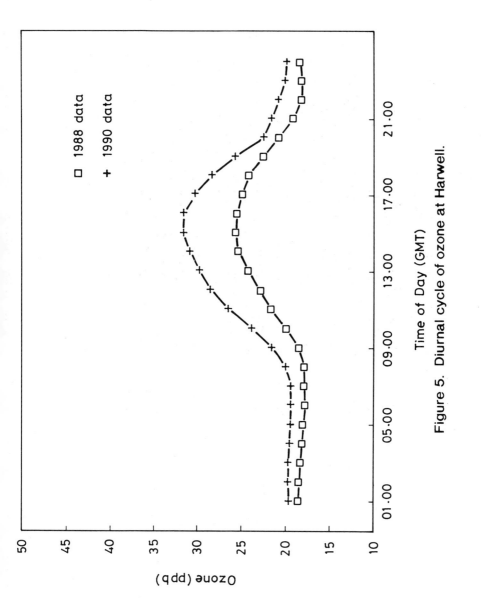

Figure 5. Diurnal cycle of ozone at Harwell.

diurnal cycle is evident in both years with the higher levels evident during 1990 which was generally a more photochemically active year. The minimum ozone concentrations occur at night time as a result of the combined effects of the lack of photochemical production and removal to the surface by dry deposition.

Figure 6 illustrates data for hydrogen peroxide measured at Harwell during 1990. There are clear diurnal patterns in concentration similar to those for ozone with night time minima and maxima occurring during the afternoon period. Figure 7 summarises the same analysis for Harwell PAN data; a behaviour very similar to ozone and H_2O_2 is evident.

Such strong diurnal patterns are not always the case. As discussed above various site features may serve to dampen the diurnal cycle of photo oxidants. Figure 8 summarises ozone data collected at the Great Dun Fell site in Cumbria. This is an elevated site (847 m amsl) and as such the site is more frequently exposed to free tropospheric air and is too high to be capped frequently by nocturnal temperature inversions hence the site is usually exposed to well mixed air and little or no diurnal variation is evident.

Variations of the nature of the diurnal cycle in ozone concentration is of importance in terms of the potential impact of ozone on vegetation. Thus, although some sites may experience large variations in ozone exposure with peak afternoon concentrations, a site such as Great Dun Fell clearly experiences a less variable ozone concentration; the interruption of nocturnal depletion processes generates a higher daily mean concentration and ozone dose to surrounding vegetation. An interesting result of such an effect is that there may be an increased exposure of vegetation to photo oxidants with increasing altitude.

From the quarterly averaged H_2O_2 and PAN data in figures 6 and 7 it is evident that there are seasonal differences in the production of photochemical species, with the strongest production during the spring and summertime periods.

In order to compare statistics or longer term trends in ozone concentrations at different measurement sites the many factors that can influence the observed ozone concentrations need to be accounted for. By considering the range in daily ozone concentrations experienced at sites and the daily mean ozone concentration it is possible to derive a site index that embodies the various factors that can affect the diurnal cycle of ozone measured at any one site. These include meterological

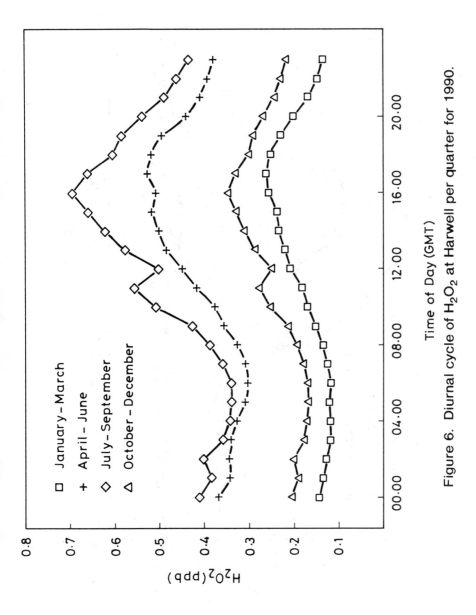

Figure 6. Diurnal cycle of H_2O_2 at Harwell per quarter for 1990.

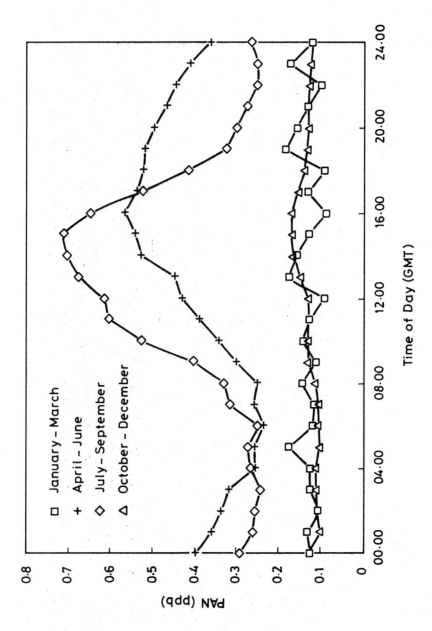

Figure 7. Diurnal cycle of PAN at Harwell per quarter for 1990.

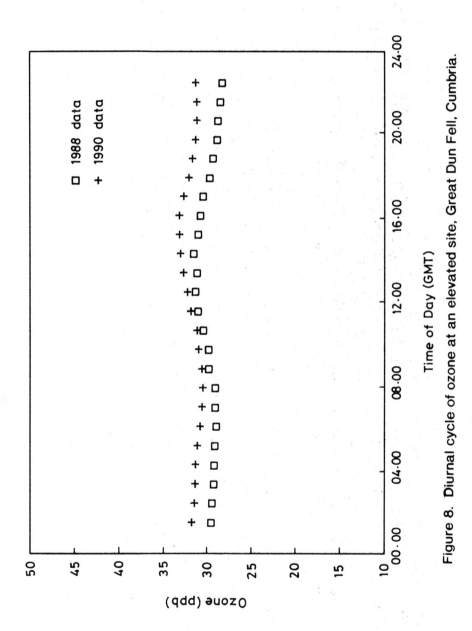

Figure 8. Diurnal cycle of ozone at an elevated site, Great Dun Fell, Cumbria.

factors, topography, arrival of polluted air at the site, the degree of night time depletion of ozone through dry deposition and the presence or absence of NO emissions (to react with ozone). An average site index has therefore been calculated by dividing the mean daily range in ozone concentration by the mean daily mean for all of the available data. Table 1 summarises the indices as an average for all years and all sites. It can be seen that the lowest ratio is exhibited by the background site at Mace Head, the two other sites classified as remote rural Great Dun Fell and Strathvaich follow on in sequence. The lower the index, the more suppressed is the diurnal cycle. The general trend is for an increasing ratio from the remote rural sites through to the London urban locations. This approach provides a useful classification of sites as it embodies the factors described above that may affect ozone concentrations and influence discernment of temporal trends in ozone data. The effect of NO_x emissions on the site index for urban sites is to enhance the index. This is due primarily to a reduction of the minimum - urban night-time ozone concentrations frequently fall to zero. Thus urban sites have a greater range; coupled with a reduced mean ozone concentration, the ratio is enhanced. It is interesting to note that Fawley and York, both classified as urban sites have lower indices than some of the sites classified as rural; prominent amongst these latter sites are those in the East Midlands, expected to be influenced by local NO_x emissions. Figure 9 is a plot of all the calculated indices for each site year against the annual mean concentration; a line fitted to the data gives an intercept on the y axis of 34 ppb ozone which may be taken to represent the annual mean concentration for a well mixed background atmosphere. The question of long term trends in ozone, H_2O_2 and PAN data is discussed below.

As mentioned above it is to be expected that because the formation of ozone is dependent, inter alia on the presence of precursor compounds and air mass quality, concentrations of photo oxidants should correlate with wind direction. Figure 10 illustrates PAN data collected at Harwell during summer 1988 and apportioned to wind direction. The enhancement of the concentrations in E and SE sectors reflects advection of polluted air from London and beyond from mainland Europe. At Harwell this tends to occur during spring and summer time antecyclonic conditions.

There are occasions when conditions for photo oxidant formation are particularly favourable and periods or episodes of enhanced photochemical activity occur. Such events can be seen in the data for all three species. The frequency with which such episodes occur varies from year to year. The occurrence of

Table 1. Listing of average site diurnal indices for PORG ozone sites

Site Name	Average Index	Type
Mace Head	0.34	RR
Strathvaich	0.43	RR
Great Dun Fell	0.58	RR
Aston Hill	0.60	R
Clatteringshaws	0.64	R
Yarner Wood	0.72	R
Wray	0.74	R
Eskdalemuir	0.77	R
Penicuik	0.79	R
Wharleycroft	0.82	R
High Muffles	0.90	R
Lullington Heath	0.90	R
Ladybower Res.	0.91	R
Lough Navar	0.92	R
North Norfolk	0.96	R
Devilla	0.99	R
Sibton	1.03	R
Harwell	1.07	R
Syda	1.10	R
Hazelrigg	1.10	R
Fawley	1.15	U
Stodday	1.18	R
York	1.27	U
St. Osyth	1.32	R
East Malling	1.33	R
Jenny Hurn	1.36	R
West Burton	1.37	R
Thorney	1.41	R
Bottesford	1.41	R
Glasgow	1.41	U
Lincoln	1.42	U
Glazebury	1.47	U
Brampton	1.54	R
Canvey Island	1.54	R
Ascot	1.61	R
Teddington	1.68	U
Stevenage	1.72	U
County Hall	1.73	U
Kew	1.75	U
Chigwell	1.76	U
St.Bartholemew's	1.80	U
Hainault	1.82	U
Nat. West Tower	1.83	U
Central London	2.04	U
Cromwell Road	2.14	U
Bridge Place	2.25	U
Islington	2.41	U

RR = Remote rural site   R = Rural site   U = Urban site

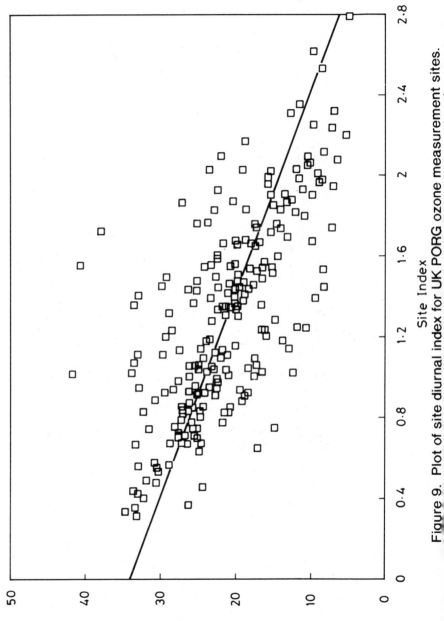

Figure 9.   Plot of site diurnal index for UK PORG ozone measurement sites.

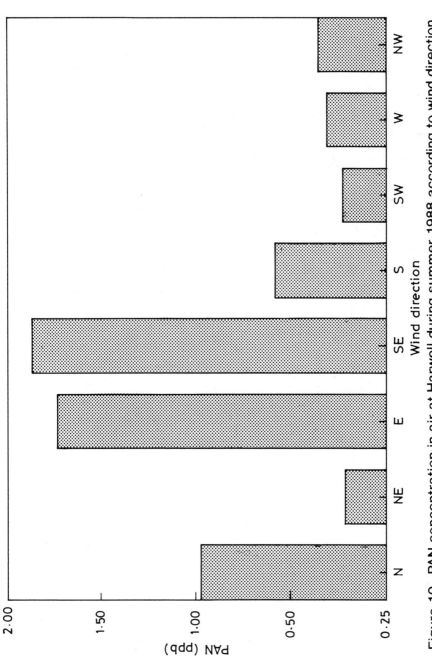

Figure 10. PAN concentration in air at Harwell during summer 1988 according to wind direction.

episodes is important in terms of human health and vegetation effects. Figure 11 illustrates ozone, PAN and H_2O_2 data recorded during an episode in 1988 at Harwell. The episode lasted 3 days with peak ozone concentrations reaching about 70 ppb. The greatest concentrations of ozone and H_2O_2 were recorded on different days; this can be explained by consideration of the origins of the air masses arriving at the site on the different days and differences in potential pollutant loading[7]. For example, on the 8th August it is likely that excess NO_x present was competing for the (HO_2) precursor of H_2O_2 serving to limit H_2O_2 production. It is interesting to note the sequencing of photo oxidant peak concentrations during the day and between days during the illustrated episode. This may be of consequence in terms of vegetation effects. Thus, for example, vegetation may be exposed to successive peak concentrations of potentially phytotoxic gases.

It has been suggested from modelling studies and some observations at background stations that northern hemispheric background ozone concentrations have increased substantially since the onset of industrialisation. More recently, since the middle of the present century, these have been increasing by up to a few percent per year.

The PORG database contains a wide range of sites some of which are influenced by NO_x emissions which as indicated earlier may serve to destroy ozone. As NO_x concentrations have been increasing during recent years this complicates the identification of longer term trends in the ozone data set and precludes the use of urban sites and the "NO_x influenced" sites with a site index of approximately 1 or more in table 1. An additional important factor is the lack of long term data records. The longest running set of measurements in the PORG database is for the Central London site (20 years). The rural site with the longest data record is Sibton. Unfortunately the Sibton data set is not complete and does appear to be a NO_x influenced site, (table 1); figure 12 is a plot of quarterly mean ozone data for Sibton. A weighted least squares analysis was used to fit a deseasonalised trend, using interpolated data for missing values. The indication is that ozone concentrations at this site have declined over the period by approximately 2% per annum.

During recent years data collection has been more systematic and more complete data sets have been compiled for several sites. It is possible to examine more recent trends using these data though by necessity this involves using short, 5 years or so, data records. Table 2 summarises the results of a weighted least

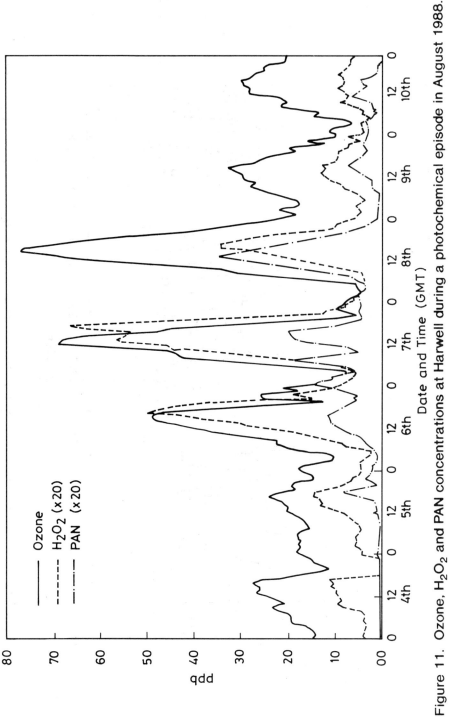

Figure 11. Ozone, H_2O_2 and PAN concentrations at Harwell during a photochemical episode in August 1988.

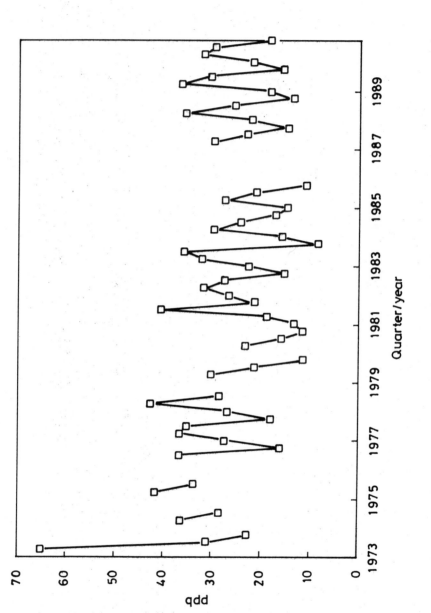

Figure 12. Quarterly mean ozone at Sibton (PORG site No.28) for the period 1973–1990.

squares analysis of monthly mean ozone trends for 3 rural sites having 5 years or more of data currently available.

Table 2. Weighted least squares regression analysis[1] of monthly mean ozone concentrations at Great Dun Fell, Lullington Heath and Aston Hill

Site Name	Height (m)	r^2	F[2]	p	Trend[3]	T[4]	p
Great Dun Fell	847	0.83	22.6	<0.001	+1.1	6.3	<0.001
Lullington Heath	120	0.96	97.7	<0.001	+1.3	8.1	<0.001
Aston Hill	370	0.75	11.9	<0.001	+1.3	4.5	<0.001

[1] SPSS (1990)[10]

[2] variance ratio regression: residuals

[3] ppb yr^{-1}

[4] T value for trend

It is evident from this subset of 3 rural sites that there has been an upward trend in ozone concentrations during the approximately five and half year period of ozone observations. Care should be taken with trend analysis performed over relatively short term data sets; however, the indication is that increases in ozone concentration can be detected at the more remote rural sites in the UK, for the 3 sites analysed this is approximately 1 ppb per year. Figure 13 summarises the ozone data for all 3 sites along with the estimated deseasonalised trend lines.

Figure 14 summarises the quarterly mean H_2O_2 and PAN data for Harwell for the period 1987-1991 together with ozone data for the same period. The notable features of the PAN data are the relatively very high concentrations recorded during April and May 1988. These values were generated during very strong photochemical episodes during this period, when data capture was relatively low. Consideration of the ozone:PAN ratios for the whole period further indicated that these data were quite different to the rest of the data set as shown in figure 15. Episodes of such strength have not been recorded since and in this respect the data is atypical of the whole series. For the analysis below these data were excluded.

Returning to figure 14 it is evident that the similar trends in the H_2O_2 and PAN data are reflected in the available ozone data. The H_2O_2 and PAN data are plotted separately as quarterly values in figure 16 where there is a very consistent

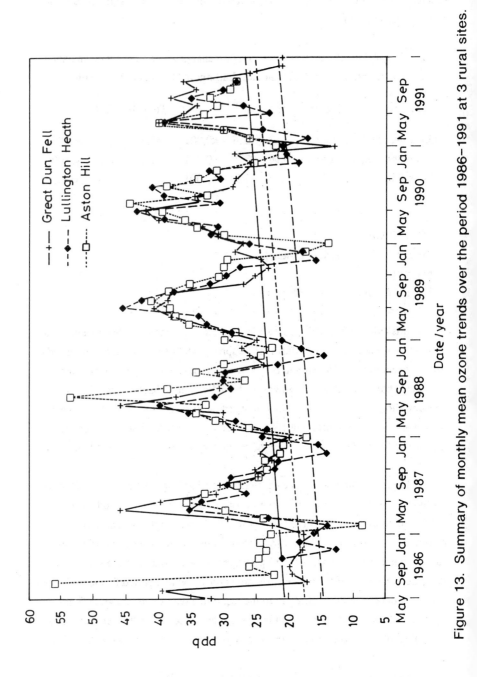

Figure 13.   Summary of monthly mean ozone trends over the period 1986–1991 at 3 rural sites.

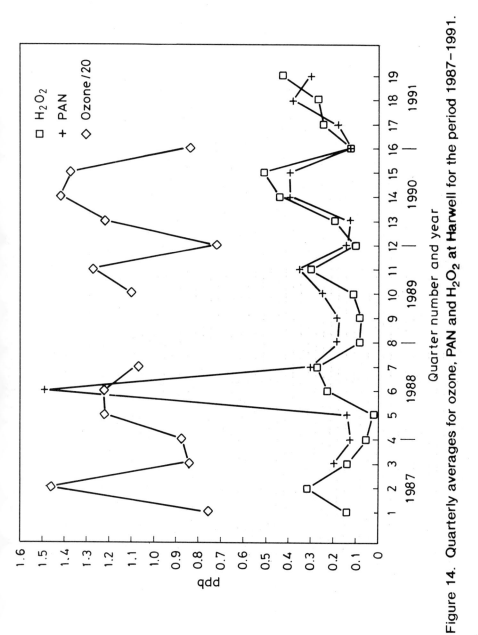

Figure 14. Quarterly averages for ozone, PAN and H₂O₂ at Harwell for the period 1987–1991.

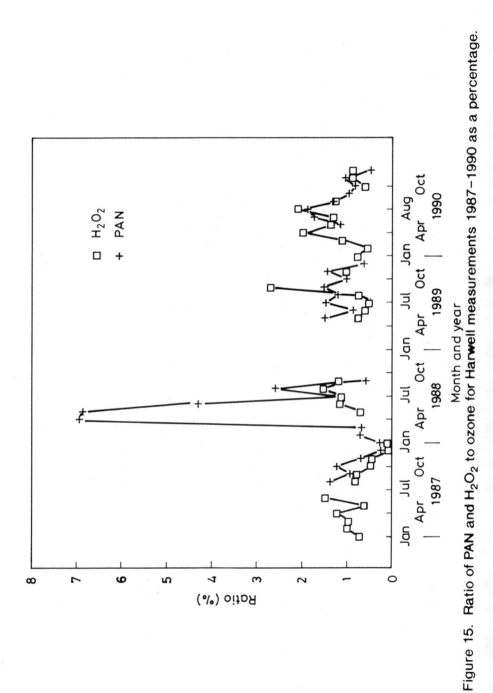

Figure 15. Ratio of PAN and H_2O_2 to ozone for Harwell measurements 1987–1990 as a percentage.

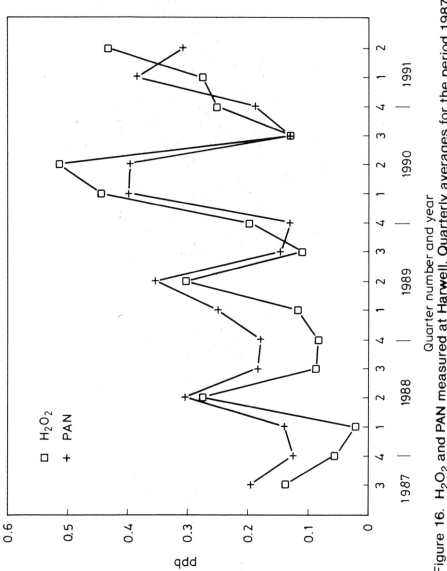

Figure 16. H₂O₂ and PAN measured at Harwell. Quarterly averages for the period 1987–1991.

pattern of behaviour between the two species. A simple regression analysis on the quarterly data gave the following relationship:

$$H_2O_2 = 1.22 \text{ PAN} + 0.08 \text{ ppb} \quad (r^2 = 0.71, n = 16)$$

The regression of ozone against H_2O_2 and ozone against PAN are shown in figures 17 and 18 respectively. The relationships indicate that for a quarterly average ozone concentration of 25 ppb one would expect about 0.3 ppb of both H_2O_2 and PAN.

Table 3.   Weighted least squares regression analysis[1] on quarterly PAN, H_2O_2 and Ozone at Harwell for the period 1987-1991

Species	n	r^2	F[2]	p	Trend[3]	T[4]	p
Ozone	20	0.57	4.6	0.014	+1.1	1.8	0.09
H_2O_2	20	0.83	17.1	<0.001	+0.028	5.3	<0.001
PAN	18	0.79	11.4	<0.001	+0.017	1.9	0.08

[1] SPSS 1990[10]
[2] variance ratio regression: residuals
[3] ppb yr^{-1}
[4] T value for trend

Table 3 summarises the results of a weighted least squares regression analysis on the data in figure 14. It may be seen that in all three cases there has been a significant upward trend in concentrations measured over the four year period.

Although the PAN and H_2O_2 data cover a relatively short period of time the evidence for increasing trends is consistent with modelling studies on future trends in photochemical oxidants[11,12], observations on PAN[13] and for H_2O_2 data collected from Greenland ice cores[14].

The PAN time series referred to above[13], recorded at Delft Holland, is summarised in figure 19, (data interpolated from figures kindly supplied by HSMA Dierden of TNO, Delft). The Delft data show an upward trend in the annual means through the 1970s with concentrations then falling to the mid 1980s. A peak in

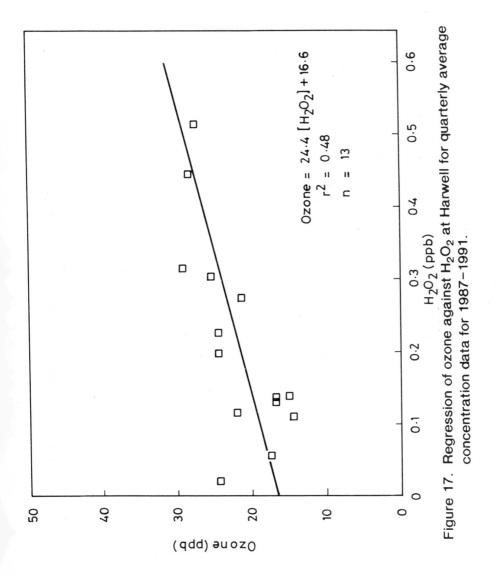

Figure 17. Regression of ozone against H$_2$O$_2$ at Harwell for quarterly average concentration data for 1987–1991.

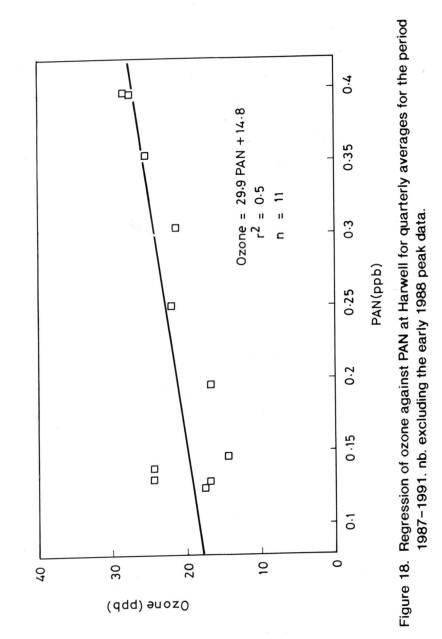

Figure 18. Regression of ozone against PAN at Harwell for quarterly averages for the period 1987–1991. nb. excluding the early 1988 peak data.

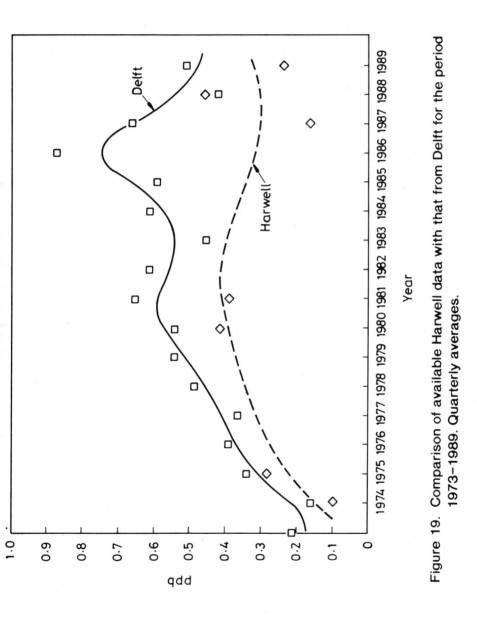

Figure 19. Comparison of available Harwell data with that from Delft for the period 1973–1989. Quarterly averages.

1986 is followed by a decline through to the end of the decade when the data record ends.

There are relatively few annual values for Harwell with which to compare the Delft trend, those available are also plotted on figure 19 with a tentative interpolated trend line. The key features of the comparison are that during the mid seventies Harwell PAN data was generally 10 to 15% lower than that in Delft, during the following 5 years PAN concentrations increased at both sites but less so at Harwell where the PAN concentrations at this time were about 25% lower than in Delft. For the data up to 1988 there appears to have been a decline in PAN concentration with the Harwell site generally registering about 40% less PAN than Delft.

The observed differences between the two sites are perhaps not unexpected given the location and nature of each, with Harwell being a rural location and Delft a more urban continental location. Evidence presented in figure 15 indicates the PAN concentrations at Harwell have been increasing during the last five years and overall the general PAN trend has been upwards since the early seventies. It is unfortunate that the Dutch time series finishes in 1989 as it is not possible to say whether the more recent data shows an increasing PAN concentration trend similar to that observed at Harwell.

Work on the time trends of photochemical oxidants in monitoring data is continuing and the UK Photochemical Oxidants Review Group will shortly publish the second report on Ozone in the United Kingdom. This will address in depth the formation, fate and consequences of photochemical oxidants in the UK.

ACKNOWLEDGEMENT

The work reported here was carried out as part of a research programme supported by Air Quality Division of the Department of Environment under contract PECD 7/12/15.

REFERENCES

1.   AMMAAPE Advisory Group on the Medical Aspects of Air Pollution Episodes. "1st Report Ozone" 1991 Dept. of Health. HMSO.

2.   PORG "Ozone in the United Kingdom" 1987. Interim report of the UK Photochemical Oxidants Review Group. Harwell Laboratory, Didcot, Oxon. OX11 0RA.

3.   S.A. Penkett, B.M.R. Jones, K.A. Brice and A.E.J. Eggleton. Atmos. Environ. 1979, 13, 123.

4.   G.P. Gervat, P.A. Clarke, A.L. Marsh, I. Teasdale, A.S. Chandler, T.W. Choulaton, M.J. Gay, M.R. Hill and T.A. Hill. Nature 1988, 333, 241.

5.   A.S. Chandler, T.W. Choularton, G.J. Dollard, M.J. Gay, M.M. Gallagher, T.A. Hill, B.M.R. Jones, S.A. Penkett, B.J. Tyler, and B. Bandy. Q. Jl. R. Met. Soc. 1989, 115, 397

6.   R.G. Derwent and P.J.A. Kay. Environ. Pollut. 1988, 55, 191.

7.   G.J. Dollard, B.M.R. Jones and T.J. Davies. Atmos. Environ, 1991, 25A, 2039.

8.   G.J. Dollard, R.G. Derwent and F.J. Sandalls. Environ. Pollut. 1989, 58, 115-24.

9.   G.J. Dollard and T.J. Davies. Environ. Pollut. 1992, 75, 45.

10.   SPSS Statistical Data Analysis SPSS/PC + Trends. 1990, SPSS Incorporated, Chicago.

11.   A.M. Hough and R.G. Derwent. Nature, 1990, 344, (6267), 645.

12.   A.M. Thompson, M.A. Owens and R.A. Stewart. Geophys. Res. Lett. 1989, 16 (1), 53.

13.   R. Guicherit. In "Tropospheric Ozone, Regional and Global Scale Interactions" edited by Isaaksen, I.S.A. D. Reidel, Dordecht, 1988, Part 1, p63.

14.   A. Sigg and A. Neftel. Nature, 1991, 351, 557.

Critical Load Concepts

M. Hornung

INSTITUTE OF TERRESTRIAL ECOLOGY, MERLEWOOD RESEARCH STATION, GRANGE-OVER-SANDS, CUMBRIA LA11 6JU, UK

1 INTRODUCTION AND BACKGROUND

The critical load concept has been developed to provide a receptor-oriented approach for use in developing emission control policies and setting emission targets. The approach determines the sensitivity of the environment to pollutant loadings and then works back to estimate the long term emissions of the pollutants that can be sustained without causing damage to the environment. The approach was first proposed in Canada in the early 1980s. It was developed further in Europe, with the Nordic countries taking the lead, through a series of workshops jointly sponsored by the Nordic Council of Ministers and the United Nations Economic Commission for Europe (UNECE)[1-3]. It is now proposed that future negotiations of SO_2 and NO_x protocols under the International Convention on Long Range Transboundary Air Pollution, under the auspices of the UNECE should be based on the critical load approach.

The definition of critical load, methods of calculation and of mapping have been refined during the workshops mentioned above. The most widely used general definition of critical load was developed and agreed at the workshop held at Skokloster in Sweden in 1988 : the critical load is "a quantitative estimate of an exposure to one or more pollutants below which significant harmful effects on specified elements of the environment do not occur according to present knowledge"[2]. The specified elements of the environment could be a whole ecosystem or a specific organism. The concept clearly implies a threshold loading,or concentration of the pollutant of interest below which there is no or insignificant damage to the ecosystem or organism being considered. In principle the critical load approach could be applied to any pollutant or combination of pollutants. However, almost all the studies to date, and the workshops noted above have focused on the determination of critical loads for acidic deposition and hence for compounds of sulphur and nitrogen.

Nitrogen has to be considered both when calculating critical loads for acidity and for nitrogen as a nutrient. Atmospheric inputs of nitrogen can lead to acidification of soils as a result of the generation of protons during nitrification of deposited NHx compounds, to produce nitrate and because nitrate in excess of plant requirements can lead to leaching of base cations. The critical load for acidic pollutants has been defined as "The highest deposition of acidifying compounds that will not cause chemical changes leading to long-term harmful effects on ecosystem structure and function according to present knowledge."[2]

Nitrogen is a limiting nutrient in many natural and semi-natural soil-plant systems but enhanced atmospheric inputs can lead to nutrient imbalances, with impacts on plant growth, changes in the competitive interactions between plants and, eventually to leaching of the excess N as nitrate. The critical load for nitrogen as a nutrient has been defined as "The maximum deposition of nitrogen compounds that will not cause eutrophication or induce any type of nutrient imbalance in any part of the ecosystem or recipients to the ecosystem."[4] Systems in which there is an apparent excess of available nitrogen above the requirements of the ecosystem have been referred to as nitrogen saturated[5].

A further important concept which has been linked to the critical load approach is that of the target load. The critical load is based on scientific criteria; it is determined by inherent properties of the ecosystem under consideration. The target load is set by a government having taken into account the critical loads for elements of the environment within the given country and also political considerations. It represents the pollutant loadings that government will seek to achieve with their emission control policies. The target load is generally less than the critical load but could be greater.

Once the critical load has been calculated for a given receptor it can be compared with data on current levels of pollutant deposition to determine whether the critical is currently exceeded. Maps showing the amount by which the critical load is exceeded in different parts of a country have come to be known as "exceedence maps".

Concepts used in the calculation of critical loads

A standard approach to the calculation of critical loads has been developed at the workshops noted above and in background documents prepared for the workshops. The first stage is the identification of the receptor system for which the critical load is to be calculated; to date, the determination and mapping of critical loads under the UNECE programme has focused on forest soils, surface waters and groundwaters as the receptors. A biological indicator is then identified —the response of which to the pollutant of interest is indicative of the response or "health" of the whole receptor system; for example a given plant species, species diversity of the vegetation, a given fish or aquatic invertebrate species. A "critical chemical value" is then set for the selected biological indicator. This value is the threshold concentration for a selected chemical determinand above which adverse changes will take place in the biological indicator. This clearly implies that a dose response relationship exists for the biological indicator-chemical determinand combination.

The brown trout has been used as the biological indicator for calculating critical loads of acidity for surface waters in north west Europe and the critical chemical value has been set in terms of alkalinity (Table 1). There is less general agreement about the most suitable biological indicator or the critical chemical value for forest soils (Table 1). The most widely used of the possible biological indicators is tree roots but the relevant data is not yet available to enable critical chemical values to be set for a number of the important natural or plantation tree species found in Britain. Relatively little progress has also been made in establishing biological indicators and related critical chemical indicators for the non-forest vegetation communities which dominate the British landscape. This reflects the leading role of the Scandinavian countries in the development of the critical load concept and the importance of forest vegetation in these countries.

Recent discussions on the critical load concept have suggested that a different biological indicator might be selected within the same receptor or ecosystem depending on the end "use" of the system. For example, if a forest ecosystem is to be protected as a complete system, retaining all its diversity and maintaining all the biological relationships, the most sensitive element of the ecosystem, or some indicator of biodiversity might be chosen as the biological target. However, if the forest is a plantation where

Receptor	Surface waters	Forest soils
Biological indicator	Brown trout	Tree roots
Critical chemical value	Zero alkalinity	Ca:Al in soil solution = 1

Table 1. Examples of receptors, biological indicators and critical chemical values being used in calculating critical loads in Europe

pH	E-layer: > pH **4.0** B-layer: > pH **4.4**
Alkalinity	> -300 µEq/l
Total aluminium	< **4.0** mg/l
Labile aluminium	< **2.0** mg/l
Ca:Al molar ratio	> **1.0**

Table 2. Critical chemical values proposed for setting critical loads for forest soils

the main objective is the production of timber, an indicator could be chosen which related more directly to the growth of the tree crop. Clearly, the most sensitive element of the system would also protect tree growth but timber production might also be maintained at levels of deposition that would adversly affect the most sensitive element of the ecosystem.

In much of north western Europe the critical load of pristine ecosystems for both acidity and N is currently exceeded and has been exceeded for some considerable time. The ecosystems have over the period of exceedence been damaged, soil and water chemistry changed and the species composition altered. It is a matter of considerable debate as to whether the critical load should be set with the aim of recreating the pristine ecosystem or some intermediate point.Such considerations would clearly affect on the choice of biological target and/or critical chemical value.

Methods used in the calculation of critical loads of acidity

Calculation of the critical load can be carried out using empirical models, simple mass balance calculations or process based computer models. The empirical models and the mass balance equations can only be used to determine critical loads when the system has reached equilibrium; they give no indication of the time to reach this equilibrium and cannot be used to determine the time taken to reach the critical load with given emission control strategies. The process based models can be used to explore these time related aspects and to evaluate the impact of a range of emission control scenarios.

An empirical model first developed in Scandanavia[6] is widely used to calculate critical loads of acidity for surface waters. This model is generally referred to as the "Steady-State Water Chemistry Method" or the Henriksen model. The critical load is calculated using the following equation;

$$CL = ([BC]_o - [ANC]_{limit}.Q - BC_d$$

CL   is the critical load for acidity

$[BC]_o$ is the pre-acidification, non-marine base cation concentration

$[ANC]_{limit}$ is the critical chemical value express in terms of acid neutralising capacity value in the drainage waters

Q is flow

BC_d is the atmospheric deposition of base cations

Further details on the use of the equation, including calculation of $[BC]_0$ can be found in Hettelingh et al[3]. All values in this and subsequent equations discussed in this paper are expressed as quantities per unit area per year, eg kg ha^{-1} yr^{-1} or mol$_c$ ha^{-1} yr^{-1}.

An alternative empirical model for use in the calculation of critical loads for surface waters has been developed by Batterbee and colleagues at University College London. This is based on results from diatom studies from a large number of lakes and uses the ratio between calcium concentration, as a measure of sensitivity, and sulphur loading. The data suggest that a ratio of 60:1 separates acidified from non acidified sites.

Critical loads for forest soils have been calculated using the "steady-state mass balance method"; and the approach can be extended to other soil types and can also be used to calculate the critical load of acidity for surface waters. The method is based on the following equation which balances all sources of alkalinity against all sources of acidity:

$$CL_{(acidity)} = BC_w - BC_u - ANC_{limit} - AC_n$$

$CL_{(acidity)}$ is the critical load of acidity

BC_w is the annual base cation release from mineral weathering

BC_u is the net long term uptake of base cations by trees or other vegetation, expressed on an annual basis

ANC_{limit} is the acceptable level of leaching of alkalinity from from the soil compartment or catchment

AC_n is the net acidity produced
                               by nitrogen uptake and
                               ammonia nitrogen uptake and
                               nitrification

The equation is usually modified to:

$$CL_{(acidity)} = BC_w - BC_u - ANC_{limit} + NH_d - NO_{totu}$$

NH_{4D} is the deposition of ammonia
                               at critical load deposition
                               levels

N_{totu} is the long term uptake of
                               nitrogen by trees and other
                               vegetation expressed on
                               an annual basis

The ANC limit is calculated from the critical chemical value for the selected biological indicator multiplied by the annual water flux from the base of the soil or from the catchment. The ANC limit is generally expressed in units of alkalinity, if the critical chemical value is, for example a pH value or a concentration of aluminium this is usually converted to units of alkalinity.

An alternative approach is to consider the critical load of potential acidity, by allowing for the potential acidifying action of ammonium in the soil[4]. The above equation then becomes:

$$CL(potential\ acidity) = BC_w - BC_u - ANC_{limit} + N_{totu}$$

An assumption of the steady state mass balance method is that the input of acidity does not further change the base saturation of the soil, the pool of adsorbed sulphate or the pool of biologically available nitrogen. It also assumes that there is no nitrogen fixation or denitrification.

A number of computer models have been developed which can be used to calculate the critical load and which effectively provide a means of solving the above equation mass balance equations, eg the PROFILE[7] and MACAL[8] models. A major problem when calculating the critical load using the mass balance method is the determination of the value for annual release of base cations from mineral weathering. It is, however a major control on the critical load. A number of approaches have been used to obtain values for given sites with the most commonly used being element budgets, based on atmospheric inputs to the site and solute outputs from the catchment or the base of the rooting zone, and detailed mineralogical studies. The PROFILE model is the only currently available model which incorporates calculation of the weathering rate; the calculation is based on mineralogical data for the site of interest and data on mineral dissolution rates obtained from laboratory studies. Most of the other models obtain a value for weathering by optimisation routines within the model or weathering rate forms part of the input data. Further research on determination of weathering rates is required given its importance in the calculation of critical loads.

A number of dynamic, process or mechanistic-based models developed to assess acidification of soils and surface waters have been adapted to calculate critical loads, eg MAGIC[9] and SMART[8]. Although differing in detail these models all incorporate the major processes controlling solid-solution interactions in soils and rocks, which control soil solution and drainage water chemistry, and a linked hydrological model of varying complexity. The main processes included are ion exchange, mineral weathering, sulphate adsorption, plant uptake of nutrients and element release from decomposition of plant residues. The models can be used to calculate soil solution chemistry and drainage water chemistry resulting from given atmospheric inputs; once the critical chemical value has been set, the models can be run to determine the level of atmospheric inputs that will not result in the value being exceeded. As noted above, the main advantage of the dynamic models is that they can be used to examine time related aspects of the critical load concept and in the scenario testing.

Determination of critical loads of sulphur and nitrogen in the context of acidification

The empirical models and the mass balance calculations considered above were developed to calculate critical loads for total acidity, ie the combined effects of sulphur and nitrogen deposition. It is clearly important, however to separate the contributions of sulphur and nitrogen when using the approach in the negotiation of protocols for control of emissions of SO_2 and NO_x. An approach has been agreed within the UNECE critical loads mapping programme[3] for calculation of the contribution of sulphur. This uses the following equation:

$$S_f = \frac{PL(SO_4)}{PL(SO_4) + PL(NO_x) + PL(NH_x) - N_u - N_{i(crit)}}$$

where S_f = sulphur fraction
$PL(SO_x)$ = present load of sulphur
$PL(NO_x)$ = present load of nitrogen
$PL(NH_x)$ = present load of ammonia
and ammonium

N_u = annual nitrogen uptake
$N_{i(crit)}$ = annual nitrogen immmobilisation in soil at
the critical load

The critical load of sulphur would therefore be:

CL(S) = S_f.CL(acidity)

CL(acidity) = critical load of acidity

The critical load of nitrogen from the point of view of its contribution to acidity could be calculated as the difference between the critical load of acidity and the sulphur fraction calculated as above. This approach has been used in an exploratory way in the UNECE critical loads mapping programme. Using this approach the critical load of nitrogen would be given by:

CL(N) = N_u + (1-S_f). CL(acidity)

CL(N) = critical load of nitrogen

N_u = nitrogen uptake

The above represents an interim, pragmatic approach and is still the subject of considerable debate. The question of the relative contributions of nitrogen and sulphur compounds in acidification will, however, become important if the concept of a "total acidity protocol", which is now being considered, is pursued. Current consideration of nitrogen is linked to the negotiation of a protocol for control of NO_x emissions but a 'total' acidity protocol would also need to consider the impact of NH_x. The estimation of the acidifying impact of SO_2 emissions is relatively straightforward but the fate of deposited nitrogen is strongly influenced by biological interactions and processes. It is important to know, for example what proportion of any deposited ammonia or ammonium is converted to nitrate, what proportion of the ammonium and of deposited NO_x are taken up by plants, what proportion is taken up via

plant leaves as opposed to plant roots. A "total
acidity protocol" would set limits for the total
emissions of potentially acidifying pollutants and ,
possibly allow some trading of S and N emissions.
Carried to its logical conclusion this would include
ammonia emissions from agriculture.

Mapping of critical loads of acidity and sulphur: progress in Great Britain

Critical load maps for acidity and for sulphur for
Europe have been produced under the auspices of the
UNECE Convention on Long-Range Transboundary Air
Pollution[3]. The maps are a composite of maps for
individual countries produced using a variety of
approaches designated level 0,1 or 2. Level 2 maps
have been derived by applying one of the available
dynamic models to a range of sites in a given country.
The level 1 maps are based on the application of the
steady state water chemistry method or the steady
state mass balance method. In level 0 maps an approach
has been used which assigns surface waters or soils to
critical load classes on the basis of their
sensitivity to acidic deposition.

The UK mapping programme for critical loads of
surface waters has used the steady state water
chemistry method[10]. A 10 x 10 km grid based on the
national grid was used and the most sensitive standing
waterbody in each 10 x 10 square was sampled. The
criteria used to select the waterbodies included areas
of highly sensitive geology[11], altitude (the highest
altitude waterbody on sensitive geology was selected),
and size (with a minimum size of 0.5 ha); waterbodies
extending between grid squares excluded. Where there
was no suitable standing waterbody in a grid square a
low order stream was sampled. The critical load for a
given waterbody, calculated based on the analysis of
the relevant water sample was assigned to the whole of
the relevant 10 x 10 km square. The resultant map
shows that surface waters with low critical loads of
acidity, <0.5 keq ha^{-1} yr^{-1} occur in northern and
western Scotland, the Lake Disrict, the north
Pennines and the Welsh uplands. The critical loads are
exceeded by current levels of deposition of acidity in
parts of these areas.

Provisional critical load maps for acidity and
sulphur have been produced for soils of Great Britain[10]
using an adaptation of an approach outlined at the
workshop held at Skokloster in 1988[2]. The workshop
suggested that soil forming materials could be
allocated to a series of classes on the basis of their
mineralogy; critical loads could then be assigned to
these classes using the weathering rates of the

relevant minerals. The underlying assumption is that base cation production from mineral weathering should be sufficient to neutralise atmospheric inputs of acidity. The development of the UK map utilised the 1:250000 soil maps and supporting databases produced by the Soil Survey and Land Research Centre and the Macauley Land Use Research Institute.The critical load map is based on the 1 km squares of the national grid. Each 1 km grid square was allocated to one of the Skokloster soil material classes, and thence assigned a critical load on the basis of the mineralogy and texture of the dominant soil series occurring in that square. This approach could not be used for peat soils as they contain little mineral material and their response to acidic deposition is not therefore controlled by mineral weathering. As a result, an approach developed by Cresser and colleagues at the University of Aberdeen was used; this is based on the change in pH which would be produced in the peat soils in response to a given input of acidity linked to a specified maximum acceptable pH change. The acidity load which would not cause a greater change in pH than the specified acceptable amount was taken as the critical load.

The resulting maps suggest that soils with low critical loads of acidity and sulphur, < 0.5 keq H ha^{-1} yr^{-1} occur in the uplands of the north and west of Britain. Overlaying of the critical load maps with maps showing current levels of deposition indicates that the critical loads are currently being exceeded over significant parts of these areas. Critical loads for a number of sites are now being calculated using the mass balance approach and dynamic models; these will then be compared with the critical loads assigned to the soils of the sites using the Skokloster based approach. A further programme of research is now in progress to validate both the surface water and soil critical load maps.

Determination of critical loads of nitrogen as a nutrient

As yet, there are no generally agreed methods for the calculation of critical loads of nitrogen as a nutrient. This partly reflects the early emphasis on sulphur and the requirements of the negotiations on the revised sulphur protocol. However, it also reflects the much greater complexity of the factors and interactions which determine the impacts of enhanced N inputs on plant nutrition and growth. Thus, for example enhanced atmospheric inputs of nitrogen can lead to a genuine excess of available N over and above the requirements of the plant and the soil microbial systems; the excess N can then be leached with impacts on soils and surface waters. Enhanced N inputs can also lead to an induced deficiency in another essential plant nutrient which is in limited supply in the particular soil-plant system and which will then lead to a limitation on the use of all the available N. The presence of large amounts of ammonium in the soil solution can block the uptake of other plant nutrients and hence result in an induced deficiency of that nutrient in the plant. Uptake of luxury amounts of nitrogen by plants above the requirements for normal plant growth can lead to increased sensitivity to water stress, frost damage and insect pests.

A number of approaches to determining critical loads for N are, however, being explored; these include the application of empirical, mass balance and dynamic modelling based methods. The empirical approach for terrestrial ecosystems uses the results of field observations and experimental additions of nitrogen compounds to determine the levels of nitrogen input at which vegetation change or a loss of vitality occur in different vegetation systems[2]. The empirically derived values quoted in[2] (op cit) are currently under revision; for example, discussions at a recent workshop in Sweden suggested critical loads of 15-20 kg N ha^{-1} yr^{-1} for lowland dry heath and 5-15 kg N ha^{-1} yr^{-1} for arctic and alpine heaths. The empirical approach applied to surface waters is based on a modification of the 'steady state water chemistry model' discussed above.

The long term mass balance approach[12, 2] is based on the following equation which seeks to balance all inputs and outputs of nitrogen in the receptor system:

$$CL_{(N)} = N_{leach} + N_{humus} + N_{biomass} + N_{dnit} - N_{fix}$$

$CL_{(N)}$ = critical load of N as a nutrient

N_{leach} = an acceptable level of N leaching below which damage is not found in the receptor system or linked ecosystems

N_{humus} = an acceptable rate of N accumulation in soil organic matter

$N_{biomass}$ = net removal of N in any crop

N_{dnit} = output of N to the atmosphere by denitrification

N_{fix} = N input from the atmosphere as a result of biological fixation

	Critical nitrogen load	
	(kg N/ha.yr)	(mmol/m^2.yr)
Heathlands	7 - 10	50 - 70
Raised bogs	5 - 10	35 - 70
Coniferous forests	10 - 12	70 - 85
Deciduous forests	< 15	< 110

Table 3. Empirical critical loads of nitrogen loads
          for natural vegetation types (after Nilsson
          and Grennfelt 1988)

Effects	Coniferous forests	Deciduous forests	Heath- lands	Sur- face waters
Vegetation changes	400-1400	600-1400	500-1400	1400
Frost damage/ Fungal diseases	1500-3000	-	-	-
Nutrient imbalances[1]	800-1500	-	-	-
Nitrate leaching	900-1500	1700-2900	2000-3600	-

[1] Refers to NH_3-N only

Table 4. Average critical loads for potential acidity
          (mol_c ha^{-1} yr^{-1}) for terrestrial and aquatic
          ecosystems in the Netherlands

          (After dr Vries and Kros 1991)

A series of dynamic models have been developed, or are under development. Some of these are based on existing hydrochemical models such as MAGIC or SAFE. Others are process-based ecosystem models such as CALLUNA and ERICA developed in the Netherlands for application to heathland ecosystems[13], or VEGIE which has been developed in the United States for application to forest ecosystems[14].

The greatest progress in determining critical loads of N as a nutrient has been made in the Netherlands and Sweden. The Dutch work has been based on a coordinated programme of research carried out under the " Dutch priority programme on acidification"[15]. A combination of field observation, experimental studies and application of models, such as CALLUNA and ERICA has enabled preliminary values of critical loads of N to be calculated for Dutch forests and heathlands[16] (Table 4).

In Sweden, a preliminary critical load map has been produced using the long term mass balance approach[17]. This suggests critical loads of c. 20 kg N ha^{-1} yr^{-1} in southern Sweden and c. 4 kgN ha^{-1} yr^{-1} in northern Sweden.

Rather less progress has been made to date in calculating and mapping critical loads of N as a nutrient for UK ecosystems. However, the UK Department of the Environment has now established an expert group to review current understanding of the Impact of Nitrogen Deposition in the Terrestrial Environment (INDITE). The group is considering impacts on vegetation, soils, and surface waters, and also aims to update quantification of N fluxes and budgets for the UK. The group will eventually apply empirical, mass balance and dynamic approaches to calculate critical loads for UK systems. Progress in the quantification of N fluxes and budgets for the UK is presented elsewhere in this volume by Fowler. The vegetation group are currently defining and mapping the distribution of N sensitive plant communities in the UK, eg Calluna and Erica dominated heath communities, ombrogenous bog communities, arctic and montane heaths and some species poor grasslands. Experimental studies are also in progress to determine critical loads for some of these communities. If one applies the empirically derived values, suggested above[2] to the UK and included in the critical loads of the lowland dry heaths and the arctic and alpine heaths, currently being exceeded over most of the area of occurrence of these plant communities in the UK.

The soil group is currently concentrating on a consideration of the concepts involved in setting critical loads for N as a nutrient and in reviewing relevant published material. However, some related work has been carried out on forest-soil systems.

Thus, studies of solute fluxes in coniferous plantations in upland Wales have indicated leaching of unexpectedly large amounts of nitrate from the older forests, > 35 years. The data could be interpreted as showing that the ecosystem is nitrogen saturated or that the critical load of N is currently exceeded. Related research suggests that deficiency of phosphorus and/or potassium, possibly induced by the increase in N deposition is effectively limiting N uptake by the trees. To date the maximum N deposition which could be used by the tree-soil system, given the limited supplies of phosphorus and potassium cannot be quantified. However, application of the empirically derived values given in Table 2 would suggest that the critical load is exceeded by the current inputs of between 15 and 25 kg N ha^{-1} yr$_{-1}$.

A series of experimental studies involving the addition of controlled amounts of N to forest systems, the NITREX project[18], has also been recently initiated in Europe with one site in north Wales. This project is designed to explore the nitrogen saturation concept and the setting of critical loads for European forest ecosystems.

The waters group is also carrying out a review of available data. The emphasis with surface waters is likely to remain on the acidifying influence of nitrogen deposition as phosphorus is generally the limiting nutrient in these systems.

References

1.   J. Nilsson, Critical loads for nitrogen and sulphur. Nordic Council of Ministers: Copenhagen, 1986.

2.   J. Nilsson & P. Grennfelt, Critical laods for sulphur and nitrogen. Nordic Council of Ministers: Copenhagen, 1988.

3.   J.P. Hettelingh, R.J. Downing, & P.A.M. de Smat, Mapping critical loads for Europe. RIVM, Wageningen. 1991.

4.   H. Sverdrup, H. de Vries & A. Henriksen, Mapping critical loads: a guidance manual to critical

calculations, data collection and mapping. Miljorapport 1990:14, Nordic Council of Ministers, Copenhagen, 1990.

5. J.D. Aber, K.J. Nadelhoffer, P Steudler & J.M. Melillo, <u>Bioscience</u>, 1989, <u>39</u>, 378.

6. A. Henriksen, L. Lien, I.S. Sevaldrud & D.F. Brakke, <u>Ambio</u>, 17, 1988, 259.

7. P Warfvinge & H Sverdrup, Water Air and Soil Pollution, 1992, 90.

8. W. de Vries, Methodologies for the assessment and mapping of critical loads and impact of abatement strategies on forest soils. Winand Staring Centre Report 46, Wageningen; the Netherlands, 1991.

9. B.J. Cosby, G.M. Hornberger & J.N. Galloway, Water Resources Research <u>21</u>, 1985, 51.

10. DOE. Acid rain - critical and target load maps for the United Kingdom, Department of the Environment: London, 1991.

11. D. G. Kinniburgh & W. M. Edmunds, The susceptibility of UK groundwaters to acid deposition. Hydrological Report, British Geological Survey No. 86/3. British Geological Survey: London, 1986.

12. P. Gundersen, <u>For. Ecol. Manage.</u>, 44, 1991, 15.

13. G.W. Heil, F. Berendse & A.H. Bakema, Effects on heathlands. In: Acidification research in the Netherlands; edited by G.J. Heij and T. Schneider. Amsterdam: Elsevier, 1991.

14. J.D. Aber, J.M. Melillo, K.J. Nadelhoffer, J. Pastor & R. Boone, <u>Ecological applications</u>, <u>1</u>, 1991, 303.

15. G.J. Heig & T. Schneider, T., Acidification research in the Netherlands. Elsevier, Amsterdam, 1991.

16. W. de Vries & J. Kros, Assessment of critical loads and the impact of deposition scenarios by steady state and dynamic soil acidification models. In: <u>Acidification research in the Netherlands</u>; edited by G.J. Heij and T. Schneider, Elsevier, Amsterdam, 1991.

17. K. Rosen,   The critical load of nitrogen to
     Swedish forest ecosystems. Department of Forest
     Soils, Swedish University Agricultural Science;
     Uppsala,   1992.

18. R.F. Wright,   Nitrogen saturation experiments
     (NITREX). Proceedings of the First European
     Symposium on Terrestrial Ecosystems, Florence 20-
     24 May 1991.   Commission of the European
     Communities, Brussels, in press.

The Chemistry and Deposition of Particulate Nitrogen-containing Species

Roy M. Harrison

INSTITUTE OF PUBLIC AND ENVIRONMENTAL HEALTH, THE UNIVERSITY OF
BIRMINGHAM, EDGBASTON, BIRMINGHAM B15 2TT, UK

1 INTRODUCTION

Atmospheric emissions of nitrogen are predominantly in the form
of gaseous compounds - oxidised nitrogen as nitric oxide, NO, and
reduced nitrogen as ammonia, NH_3. In the atmosphere these
species are converted to particulate compounds - nitric oxide by
oxidation first to nitrogen dioxide and then nitric acid, and
subsequent reaction with ammonia or existing aerosol such as
sodium chloride. Ammonia gas reacts predominantly in acid-base
processes to form ammonium salts. These particulate species are
in the main comprised of small particle sizes and deposit
predominantly in wet deposition, rather than dry processes which
are relatively inefficient. In this article, formation routes for
particulate species of nitrogen will be examined, their chemistry
considered and their role in deposition of nitrogen to land and
waters assessed.

Formation Routes and Chemistry of Particulate Nitrogen Species

Emissions of oxidised nitrogen are mainly in the form of nitric
oxide, although many sources emit also a few percent of nitrogen
dioxide. After dilution with ambient air, a photostationary state
equilibrium is set up involving NO, NO_2 and O_3[1]. Under most
circumstances, remote from major sources, NO_2 predominates.
This compound is oxidised to nitric acid, HNO_3, present in the
atmosphere as a vapour. During daytime the main oxidation
mechanism is by reaction of NO_2 with the hydroxyl radical, OH

$$NO_2 + OH + M \longrightarrow HNO_3 + M \qquad\qquad (1)$$

The atmospheric lifetime of NO_2 with respect to this process is typically around 12 hours. At nighttime photolytically-generated OH is absent and oxidation proceeds through other mechanisms involving the nitrate radical, NO_3. The rate of production of HNO_3 <u>via</u> these processes is uncertain, but it appears that considerable nitric acid can arise through this route. The first stage is oxidation of NO_2 by O_3

$$NO_2 + O_3 \longrightarrow NO_3 + O_2 \qquad\qquad (2)$$

This can be followed by direct abstraction of a hydrogen atom from an aldehyde or hydrocarbon

$$RH + NO_3 \longrightarrow R + HNO_3 \qquad\qquad (3)$$

Alternatively NO_3 may react reversibly with further NO_2 to form N_2O_5 which reacts on aqueous surfaces to give HNO_3

$$NO_3 + NO_2 \rightleftharpoons N_2O_5 \qquad\qquad (4)$$

$$N_2O_5 + H_2O \longrightarrow 2HNO_3 \qquad\qquad (5)$$

Once formed, nitric acid is subject to efficient dry deposition, or to reaction with ammonia gas in a reversible process which forms particulate ammonium nitrate

$$HNO_3 + NH_3 \rightleftharpoons NH_4NO_3 \qquad\qquad (6)$$

In unpolluted regions insufficient of the precursor gases is typically present to cause formation of ammonium nitrate aerosol. In polluted areas, however, appreciable formation occurs. Seinfeld and coworkers[2,3] have used equilibrium chemical thermodynamics to calculate the concentration product $[HNO_3][NH_3]$ in equilibrium with NH_4NO_3. This is a strong function of both temperature and relative humidity. Above the deliquescence point of NH_4NO_3, liquid droplets form and equilibrium concentrations of HNO_3 and NH_3 in contact with the droplets are lower than for solid particles.

Another contributor to ammonium salt aerosol is ammonium chloride[4]. In this instance the hydrochloric acid is derived primarily from coal combustion and refuse incineration[5]. Again, the formation reaction is reversible

$$NH_3 + HCl \rightleftharpoons NH_4Cl \tag{7}$$

Equilibrium concentration products $[HCl][NH_3]$ as a function of relative humidity and temperature have been calculated by Pio and Harrison[6,7].

The dissociations of NH_4NO_3 and NH_4Cl are amongst rather few examples of atmospheric chemical processes which may be subject to thermodynamic rather than kinetic control. There has been considerable interest in investigating the extent to which these chemical systems do conform in the atmosphere to the predictions of chemical thermodynamics. This is exemplified by Figure 1, showing the work of Allen and coworkers[8] in the U.K. The theoretical concentration products are shown as solid lines for different relative humidities. The uncertainties for solid NH_4NO_3 relate to different sources of thermodynamic data. The measurement data points are generally appreciably above the theoretical lines. This behaviour was found irrespective of measurement method and averaging time (3 h or 24 h) and occurred also in measurements made in the Netherlands[8]. Both NH_4NO_3 and NH_4Cl showed comparable behaviour. The greatest divergences from thermodynamic equilibrium occurred at low temperatures and high relative humidities. Similar behaviour in terms of large positive deviations of measured concentration products from predicted thermodynamic equilibrium has occurred in most other studies, with the exception of measurements in Sweden where concentration products were frequently too low to sustain formation of NH_4NO_3 aerosol[9].

The deviations from predicted thermodynamic equilibrium are not explicable in terms of an internally mixed, as opposed to externally mixed aerosol (in the former, individual particles contain more than one compound; in the latter each particle contains only one compound, possibly with water also). In an internal aerosol mixture, equilibrium concentration products would be lower than for the external mixture.

In attempting to explain their results, Allen et al.[8] suggested that kinetic constraints limited the rate at which NH_3 and HNO_3 were incorporated into the aerosol. This was pursued further by Harrison et al.[10] who gained a measure of the reaction rates by determining the rates of evaporation of both dry and aqueous particles of NH_4Cl and NH_4NO_3 in the laboratory. The evaporation rate, expressed as loss of particle radius per unit time, was found to be independent of radius contrary to the expected dependence

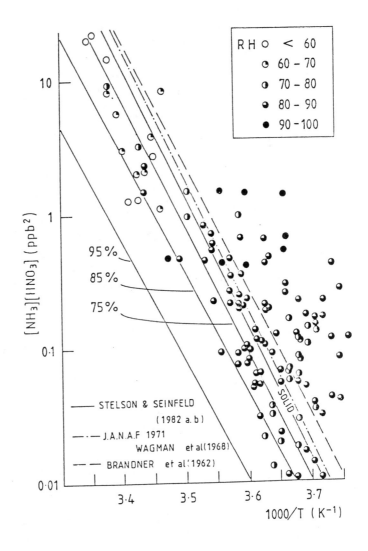

<u>FIGURE 1</u>   Measured concentration products (points) and
                  theoretical equilibrium values (lines) as a
                  function of temperature and relative humidity. fc
                  3 hour samples of nitric acid and ammonia
                  collected in Eastern England.

upon (radius)$^{-1}$ for a gaseous diffusion-limited process. These results were further explored by Harrison and MacKenzie[9] who analysed the data for both formation and evaporation reactions of NH_4NO_3 and NH_4Cl, concluding that chemical kinetic limitations determined the rates. Rate data were incorporated in a numerical model of the chemistry leading to formation of NH_4NO_3 in the atmosphere, which predicted disequilibria of a similar magnitude to those measured in the field[9].

A wholly theoretical treatment of the kinetics of the $NH_4NO_3/HNO_3/NH_3/H_2O$ and $NH_4Cl/HCl/NH_3/H_2O$ systems by Wexler and Seinfeld[11] came to broadly similar conclusions. In this work, characteristic times for the achievement of equilibrium were estimated. These varied from seconds to hours. The slowest equilibration occurred under low temperature, high humidity conditions with low particle loadings, consistent with the field measurements of Allen et al.[8]

The other strong acid typically present at appreciable concentrations in polluted atmospheres is sulphuric acid. This is formed from sulphur dioxide oxidation and exists in the atmosphere as fine, often sub-micrometre, aerosol. Reaction with ammonia occurs in two stages

$$NH_3 + H_2SO_4 \longrightarrow NH_4HSO_4 \qquad (8)$$

$$NH_4HSO_4 + NH_3 \longrightarrow (NH_4)_2SO_4 \qquad (9)$$

The rate of these processes has been very difficult to determine, and indeed it is not possible to discriminate between the two reactions in the atmosphere, as a continuum exists of progressive neutralisation of acidity. Laboratory studies with fresh H_2SO_4 droplets have shown that ammonia molecules colliding with the surface have an almost 100% probability of reaction and thus except for very small particles (< 0.1 μm diameter) the reaction is gaseous diffusion limited. In practice other constraints may be important as much H_2SO_4 is formed within clouds, or at cloud altitudes, whilst NH_3 is released at the ground surface. Thus concentrations of H_2SO_4 are lowest where NH_3 is most abundant and *vice versa*, and consequently atmospheric mixing processes have a strong influence upon the reaction rate.

There are very few measurements of the rate of the acid sulphate/NH_3 reaction. A rate constant, k, defined by

$$\frac{-d}{dt} [\text{aerosol } H^+] = k[NH_3] \qquad\qquad (10)$$

has been estimated by Harrison and Kitto[12] from horizontal gradients in concentration of aerosol acidity in air masses advected over ammonia-rich land from the ammonia-depleted North Sea atmosphere. Values of $(0.04 - 4) \times 10^{-4}$ s^{-1} were measured, consistent with estimates by Erisman et al.[13] based upon vertical gradients in NH_3 and aerosol H^+.

Once formed, NH_4HSO_4 and $(NH_4)_2SO_4$ are relatively involatile and do not dissociate appreciably to their precursors in the atmosphere.

2 DEPOSITION OF PARTICULATE NITROGEN

Deposition velocities[13] for particulate materials are crucially dependent upon particle size. A minimum occurs in the deposition velocity in the accumulation size range of ca 0.1 - 1 μm aerodynamic diameter. Some typical particle size distributions of major ion species in polluted air masses sampled in northern England[14] appear in Figure 2. In relation to nitrogen species, these show:

(i) ammonium is unimodal with a modal diameter of 0.3 - 0.5 μm, closely correspondent to that for sulphate

(ii) nitrate is bimodal with a fine particle mode corresponding to that for ammonium and sulphate, and a coarser mode at ca 2 μm. This latter mode is seen at larger sizes in samples containing a greater proportion of marine-derived aerosol.

Dry deposition velocities are consequently rather low, but usually higher for nitrate than for ammonium.

Fluxes to Land

Land Surface exchange of the gaseous species, HNO_3 and NH_3 can be evaluated by the micrometeorological gradient method. Were the association/dissociation reactions of NH_4NO_3 and NH_4Cl rapid, they would render this method unreliable as the species would not be conserved, but would change chemically. Work in the U.K. by Harrison et al.[15] indicates that in this situation the equilibration reactions are insufficiently fast to influence the

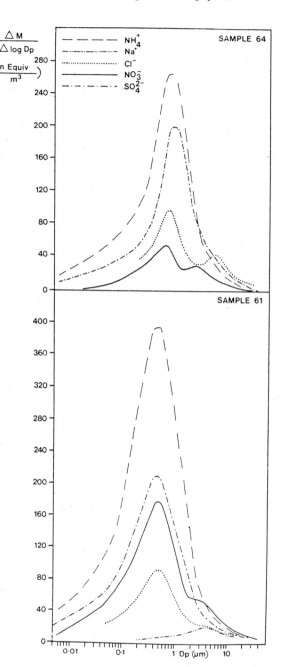

FIGURE 2   Size distributions of major ion species in two representative airborne particle samples collected in Northern England

process. Work in the United States has, however, drawn the opposite conclusion[16,17].

Harrison and Allen[18] have used data from sites in eastern England to estimate annual deposition of nitrogen in various forms. The results are summaried in Table 1. For the particulate species, NH_4^+ and NO_3^-, wet deposition is far more important than dry (although there may be a contribution to wet deposited NO_3^- and NH_4^+ from scavenging of HNO_3 and NH_3). The dominant term in this budget is ammonia dry deposition. The method of calculation assumed that land was untreated by chemical fertilisers and animal manure. In this situation, research shows a substantial downward transfer of ammonia from the atmosphere under most circumstances. In the case of fertilised land or that used for keeping animals, net fluxes are frequently upwards from the ground to the atmosphere.

Table 1   Estimated Annual Deposits of Nitrogen in Eastern England in Various Chemical Forms[18]

	Assumed v_g (mm s^{-1})	Flux (kg N ha^{-1} a^{-1})
Wet Deposition		
NO_3^-- N		3.2
NH_4^+- N		3.9
Dry Deposition		
NO_3^-- N	1.5	0.7
NH_4^+- N	1.5	0.9
HNO_3- N	22	1.6
NH_3- N	22	8.6
NO_2 - N	1.0	2.8

Fluxes to the Sea

Recent work has sought to refine estimates of atmospheric deposition of nitrogen to the North Sea. This has involved measurement of concentration fields of HNO_3, NH_3 and aerosol NH_4^+ and NO_3^- over the southern bight of the North Sea[19] and measurements of precipitation composition and depth[20]. The concentration field of aerosol nitrate is shown in Figure 3. Table 2 shows atmospheric deposition of nitrogen species to the

<u>Table 2</u> Estimated Annual Atmospheric Deposition
of Nitrogen Species[21] to the southern North Sea
(south of 56°N), 10^3 t N a[-1]

	HNO_3-N	NO_3^--N	NH_3-N*	NH_4^+-N	DON-N†	Total-N
Dry deposition	20.0	44.7	14.9	22.2	-	101.8
Wet deposition	-	70.5	-	47.8	8.1	126.4
Total deposition	20.0	115.2	14.9	70.0	8.1	228.2

*   Value for NH_3 is an upper limit
†   DON is dissolved organic nitrogen

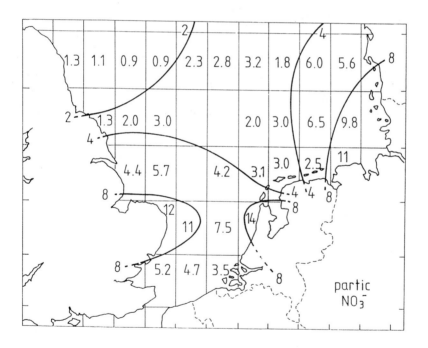

<u>FIGURE 3</u>   Average airborne concentrations of particulate
nitrate in air over the North Sea, by grid square.

southern North Sea (south of 56°N) as a function of species and mode of deposition[21]. In this case wet deposition of NO_3^- and NH_4^+ is around double the dry deposition of these species. Particles sizes of nitrate measured over the North Sea were found to be far larger than for ammonium[19]. If appears that nitric acid vapour is reacting with large sodium chloride aerosol

$$NaCl + HNO_3 \longrightarrow NaNO_3 + HCl \qquad\qquad (11)$$

The resultant $NaNO_3$ is of large particle size and higher deposition velocity than the ammonium, much of which is in the accumulation mode. Concentration products of $[HNO_3][NH_3]$ are typically below the predicted thermodynamic equilibrium values and it appears that NH_4NO_3 evaporates in the low-ammonia atmosphere over the sea.

3 CONCLUSIONS

Chemical transformations influence greatly the chemical make-up of reactive nitrogen in the troposphere. Oxidation reactions and acid-base processes are of particular importance. The particulate species are more subject to wet than dry deposition and make a significant contribution to total nitrogen fluxes to both land and sea surfaces.

REFERENCES

1. R.M. Harrison, 'Pollution: Causes, Effects and Control', R.M. Harrison (ed.), Royal Society of Chemistry, London, 1990, Chapter 8, p. 157.

2. A.W. Stelson and J.H. Seinfeld, Atmos. Environ., 1982, 16, 983.

3. A.W. Stelson and J.H. Seinfeld, Atmos. Environ., 1982, 16, 993.

4. L.A. Barrie, R.M. Harrison and W.T. Sturges, Atmos. Environ., 1989, 23, 1083.

5. P.J. Lightowlers and J.N. Cape, Atmos. Environ., 1988, 22, 7.

6. C.A. Pio and R.M. Harrison, Atmos. Environ., 1987, 21, 1243.

7. C.A. Pio and R.M. Harrison, Atmos. Environ., 1987, 21, 2711.

8. A.G. Allen, R.M. Harrison and J.W. Erisman, <u>Atmos. Environ.</u>, 1 989, <u>23</u>, 1591.

9. R.M. Harrison and A.R. MacKenzie, <u>Atmos. Environ.</u>, 1990, <u>24A</u>, 91.

10. R.M. Harrison, W.T. Sturges, A.M.-N. Kitto and Y. Li, <u>Atmos. Environ.</u>, 1990, <u>24A</u>, 1883.

11. A.S. Wexler and J.H. Seinfeld, <u>Atmos. Environ.</u>, 1990, <u>24A</u>, 1231.

12. R.M. Harrison and A.M.-N. Kitto, <u>J. Atmos. Chem.</u>, in press.

13. R.M. Harrison in 'Pollution: Causes, Effects and Control', R.M. Harrison (ed.), Royal Society of Chemistry, London, 1990, Chapter 7, p. 127.

14. R.M. Harrison and C.A. Pio, <u>Atmos. Environ.</u>, 1983, <u>17</u>, 1733.

15. R.M. Harrison, S. Rapsomanikis and A. Turnbull, <u>Atmos. Environ.</u>, 1989, <u>23</u>, 1795.

16. B.J. Huebert, W.T. Luke, A.C. Delany and R.A. Brost, <u>J. Geophys. Res.</u>, 1988, <u>93</u>, 7127.

17. R.A. Brost, A.C. Delany and B.J. Huebert, <u>J. Geophys. Res.</u>, 1988, <u>93</u>, 7137.

18. R.M. Harrison and A.G. Allen, <u>Atmos. Environ.</u>, 1991, <u>25A</u>, 1719.

19. C.J. Ottley and R.M. Harrison, <u>Atmos. Environ.</u>, in press.

20. T. Jickells and A. Rendell, unpublished data.

21. A. Rendell, C.J. Ottley, T.D. Jickells and R.M. Harrison, <u>Tellus</u>, submitted.

Wet Deposition

J. G. Irwin, G. W. Campbell, and J. R. Stedman

WARREN SPRING LABORATORY, GUNNELS WOOD ROAD, STEVENAGE, HERTFORDSHIRE SGI 2BX, UK

1 INTRODUCTION

Wet deposition is a major pathway for the transfer of nitrogen species to the earth's surface. In recent years the availability of data from networks of precipitation collectors has allowed concentration and deposition maps of nitrate and ammonium to be drawn on regional and national scales. In this paper uncertainties in precipitation sampling are first briefly reviewed. In mountainous regions a substantial fraction of precipitation originates from the scavenging of cap cloud. This results in both increased rainfall and larger pollutant concentrations compared to surrounding lowland areas. The importance of this mechanism over the United Kingdom is discussed and quantified in order to estimate total wet deposited nitrogen.

Few long-term data sets for nitrate and ammonium in rain are available but measurements have been made to a consistent protocol for eleven years at a site in southern Scotland. These data allow changes in sulphate and nitrate to be examined in the context of changing emissions of sulphur dioxide and oxides of nitrogen.

2 PRECIPITATION SAMPLING

Despite its apparent simplicity, precipitation sampling is loaded with a number of systematic errors which are imperfectly understood. These errors can cause considerable losses in precipitation samples, typically amounting to 5-10% for rain and up to 50% or more for snow. The two most important of these are evaporative losses and sampling losses due to wind effects; the discussion here concerns the latter.

The majority of precipitation collectors are containers or funnels of some sort which are set on or above the ground and are thus exposed to the wind. Virtually all have a substantial bulk and the resulting aerodynamic blockage causes a rising and accelerating airflow over the opening of the gauge. This causes precipitation to be displaced away from the opening with some loss of sample[1,2]. The problem increases in severity as the wind speed increases and as the precipitation falling speed is reduced. Thus it is most significant for snow and fine rain and at exposed sites where average windspeeds are high. There is an additional effect which can also cause sample loss; this is the wind-driven recirculating airflow inside the collector. This is capable of discharging precipitation which has already entered and of enhancing evaporative losses.

A quite effective solution to the problem is to bury the collector so that its opening is level with the ground, thereby removing wind effects. The collector is normally set in a pit, surrounded by a grid or specially prepared surface to avoid splashing[3]. However, this solution is not practicable in many cases. Moreover, it is not acceptable for sampling snow, or where samples are to be analysed chemically, since unwanted material blown along the ground is also collected.

Research is now underway, using both wind tunnels and computational fluid dynamics to develop more aerodynamic collectors[4]. At present, however, the great majority of precipitation collectors are mounted above the ground and suffer from the wind induced sampling losses described above. In the United Kingdom two types of collector are used in the national monitoring programme. Bulk collectors[5], which are continually open, are used to collect weekly samples at 32 sites. At five of these sites wet-only collectors, which open only when it is raining, are also installed. Comparison with Meteorological Office standard gauges has shown losses to range from 1 to 23% for bulk collectors and 10 to 35% for wet-only collectors, the lower collection efficiencies being observed at exposed sites experiencing high wind speeds[6]. Figure 1 illustrates this point.

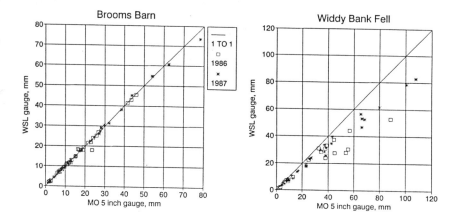

Figure 1   Comparison of weekly rainfall measured by the WSL bulk precipitation collector and Meteorological Office standard rain gauge at Brooms Barn and Widdy Bank Fell.

At 510m Widdy Bank Fell is the highest site in the network and is located on exposed moorland. In contrast Broom's Barn is a lowland site, rarely experiencing wind speeds greater than 6ms^{-1}. An additional problem which arises with wet-only collectors is the requirement to detect accurately the onset of precipitation. In general, most sensors detect precipitation at or above 1mmh^{-1} fairly well; typically 98% of precipitation falls with an intensity greater than this and, hence, in most cases precipitation sensing is not a major source of error.

This undercatch, which is similar to the performance of other collectors used elsewhere in the world, may be accounted for by using more accurate rainfall data to calculate deposition fields. The question remains, however, as to whether the chemical composition of the sample is representative of the precipitation which fell. As small raindrops tend to be the most polluted[7] and also the least efficiently collected, it must be assumed that the results are biassed to some degree. In the absence of a perfect collector this is difficult to assess, but comparisons between collectors of differing collection efficiencies suggest that, in practice, the effect is comparatively small.

A further question concerns dry deposition to bulk collectors. Wet-only collectors are intended to reduce contamination of precipitation samples by dry deposition of gases and particles. Those used in the national network include a refrigerated cabinet to store the collected rainwater at 4°C. In general agreement between the two types of collector for ionic composition under United Kingdom conditions is good[8]. At two sites which experience low rainfall and elevated concentrations of primary and secondary pollutants in easterly airflow the bulk collector has been found to give larger concentrations of non-marine sulphate; measurements of particulate sulphate suggest that dry deposition of particles is sufficient to explain the differences. In contrast, larger ammonium concentrations are measured with the wet-only collector. The observations are consistent with loss of ammonium from the bulk collector, particularly at sites with low annual mean acidity. These differences are greater in summer, consistent with a biological loss mechanism. While wet-only collectors are preferable it has been concluded[9] that, under United Kingdom conditions, carefully sited and operated bulk collectors provide valid information on spatial patterns.

3 MAPPING

The technique of kriging has been used to map precipitation-weighted nitrate and ammonium concentrations. The procedure has been described elsewhere[10]. One advantage of kriging is that the interpolation variance is known for each interpolated estimate and can be mapped along with the concentration to provide a measure of the reliability of the map. Before considering the spatial distributions it should be remembered that the network was designed to characterise regional precipitation chemistry as opposed to that influenced by urban areas or local pollutant sources. The perturbation of these regional patterns in an urban area is considered later.

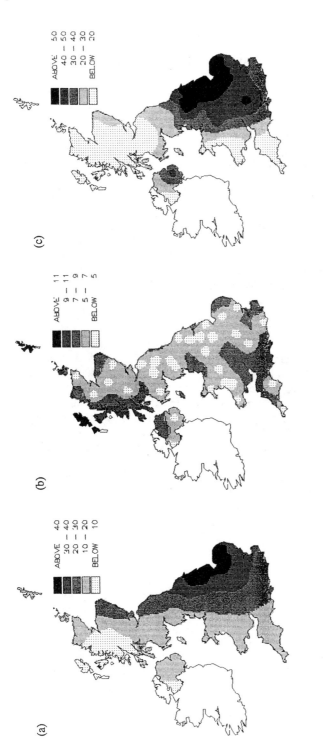

Figure 2   Maps showing (a) precipitation-weighted mean nitrate concentration, (b) the associated interpolation error and (c) precipitation-weighted mean ammonium concentration in 1990 (μeql^{-1}).

Concentration

The concentration of nitrate during 1990 is mapped in Figure 2a. Nitrate concentrations ranged from less than 10 μeql^{-1} in north-west Scotland and the west of Northern Ireland to greater than 40μeql^{-1} in eastern England and East Anglia. The smallest and largest concentrations being 7 and 46 μeql^{-1} at sites in Fermanagh and Lincolnshire respectively. The associated error map is presented as the kriging error (Figure 2b). The residuals are normally distributed and thus 95% lie within twice the error from the mean. The error is dependent on the distance of the estimated point from measurement sites and ranges from less than 5 μeql^{-1} to greater than 11μeql^{-1}. Over most of the country it is less than 7μeql^{-1}, corresponding to less than 25% over most of the high concentration areas.

Ammonium concentrations, shown in Figure 2c, were also smallest in the north and west, where concentrations were less than 20μeql^{-1}, increasing to over 50 μeql^{-1} in parts of eastern England. The largest annual mean concentration, 69 μeql^{-1} was recorded in Norfolk and the lowest, 4μeql^{-1} in northern Scotland. These spatial patterns are consistent with earlier years[11,12].

Deposition

In effects studies it is often the deposition rather than the concentration of pollutant species which is of greatest interest. Figures 3a and b show the deposition of nitrate and ammonium in 1990. These maps have been calculated by combining the interpolated concentration fields with interpolated rainfall rather than using measurements of deposition directly. The advantage of this approach is that it makes use of the dense network of rain gauges co-ordinated by the Meteorological Office. Implicit in this approach is the assumption that rainfall composition does not vary with altitude; this is not the case and the modification of these estimates in mountainous areas is discussed in the following section. For both species the areas of largest deposition fall into two classes: Those close to major source areas, for example parts of East Anglia, and those areas more remote from sources but experiencing high rainfall such as the west Central Highlands of Scotland, Galloway and Cumbria. In these regions both nitrate and ammonium annual deposition may exceed 0.5 gNm^{-2}.

Seeder-feeder enhancement

As the collectors used to define spatial patterns are largely confined to low altitude sites for practical reasons it is implicit in the mapping procedure that precipitation composition is constant with altitude. As noted above, this is not in fact the case. In mountainous areas a substantial proportion of rainfall arises from scavenging of cap cloud by the seeder-feeder process (Figure 4a). In this process, aerosol containing nitrate, ammonium and other ions is lifted by the hills and activated into orographic cloud with droplet radius typically in the range 3 to 15 μm. These droplets are efficiently scavenged by precipitation falling from higher levels, whereas the unactivated aerosol upwind of the high ground is mainly of diameter less than 1 μm and is much less efficiently scavenged. The cap cloud droplets contain larger concentrations, possibly an order of magnitude or more,

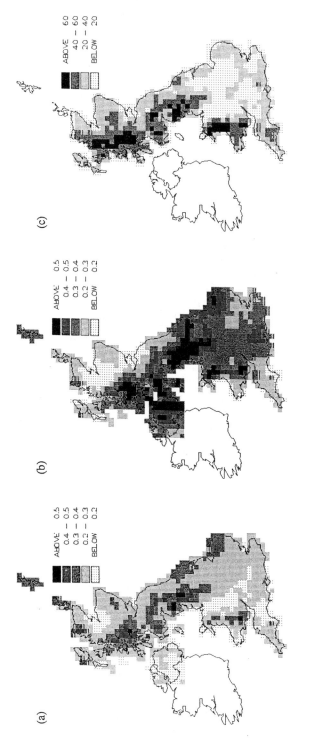

Figure 3   Maps showing (a) nitrate deposition and (b) ammonium deposition in 1990 (gNm^{-2}); and (c) percentage enhancement of nitrate deposition due to seeder-feeder scavenging.

than the rain from higher levels. Research at Great Dun Fell in Cumbria[13] has shown that concentrations of major ions in rain increased with altitude by between a factor of 2 and 3 over the range 200 to 850 m. Sulphate, chloride and ammonium behaved in a similar way to nitrate (illustrated in Figure 4b). As the amount of rain also approximately doubled over this height range, wet deposition increased by between a factor of 4 and 6.

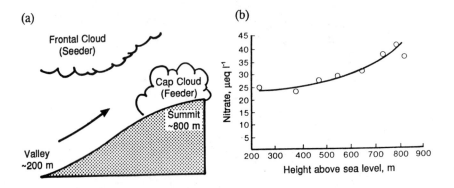

Figure 4   (a) Seeder-feeder mechanism for enhanced rainfall concentrations of ammonium and nitrate and (b) variation of nitrate concentration with altitude at Great Dun Fell.

These findings are supported by other similar studies[14,15] and also by investigations of ^{210}Pb deposition in south-west Scotland which imply an approximate doubling of concentration in the feeder cloud. The absence of detailed studies throughout the country inevitably makes extrapolation to a national scale somewhat uncertain. However, orographic enhancement over the country has been calculated on the basis of the concentrations in feeder cloudwater exceeding those in seeder rain by a factor of two. In the more polluted Pennines, this value may be conservative. In the least polluted parts of western Scotland and Ireland, feeder cloud water concentrations may be smaller.

Figure 3c shows the percentage enhancement for nitrate deposition. Over much of the country the effect is small but in some mountainous areas it can increase deposition by as much as 70% over 20x20 km grid squares and even more over isolated peaks. Many of the areas experiencing large seeder-feeder enhancements are in sensitive parts of the country where critical loads are small. Hence, while the effect is small in terms of the national budget it does have a disproportionately large influence on biological effects in sensitive ecosystems.

Urban concentrations

A major study of precipitation chemistry has been carried out in Manchester[16]. The network of sites installed in the conurbation was designed to allow maximum comparability with the national network. Bulk collectors of the same design were used and similar sampling protocols adopted. Siting criteria followed those used nationally although inevitably the urban environment resulted in some compromise. Maps for 1989 are shown in Figure 5. In general concentrations are consistent with the regional patterns which indicate concentrations in the ranges 20 to 30 μeql^{-1} and 30 to 40 μeql^{-1} for nitrate and ammonium respectively. The measurements do, however, reveal the scale of variability within an urban area. This is in agreement with other studies which have found little evidence of increased nitrate concentrations[17] but some limited evidence, particularly in London, of elevated ammonium concentrations[18,19].

(a)   (b)

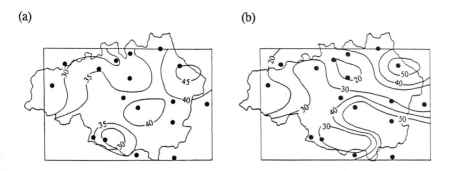

Figure 5   Concentrations of (a) nitrate and (b) ammonium (μeql^{-1}) over the
Greater Manchester area in 1989 (reproduced from reference 16).

4 TRENDS

Seasonal variations in concentration and deposition are also relevant to effects studies. All major ions show a marked seasonal variation in concentration (Figure 6) with largest concentrations in late-spring and early summer, while rainfall exhibits an approximately inverse variation, resulting in a smaller seasonal variation in deposition. Although concentrations are markedly larger in the spring, the spatial pattern varies little during the year[11]. More detailed examination has, as yet, failed to reveal any significant differences in the seasonal variations between nitrate and ammonium ions. Current knowledge suggests that precipitation amount and other meteorological factors are largely responsible for the seasonal cycle in concentration. In contrast, concentrations of marine-derived ions, such as sodium, are largest in winter due to higher wind speeds and the frequency of occurrence of maritime air masses. Deposition of marine ions is, therefore, also largest in winter.

Few long time-series of data of known quality exist; measurements at Rothamsted suggest that nitrate concentrations may have doubled since the beginning of the century[20]. The time-series for Eskdalemuir in southern Scotland shows that while sulphate concentrations have decreased, those of nitrate and ammonium have remained effectively constant (Figure 7).

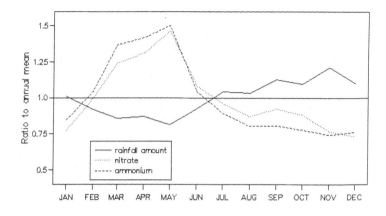

Figure 6   Seasonal variation in nitrate and ammonium concentration averaged over
            32 sites.

This is clearly illustrated in the sulphate to nitrate ratio, which parallels the
relative change in emissions. The daily measurements at Eskdalemuir allow
interpretation in terms of the back-trajectories of air masses arriving at the site;
these are used to assign each day to a 45° transport sector. In some cases this is
not possible and the day is assigned to a ninth, indeterminate, sector[11]. Although
there is greater noise, a downward trend is evident in all the sectors for which
adequate data (more than 20 measurements in each 3-year period) are available
(Figure 8). This decrease is most marked in sector 8 which includes the Glasgow

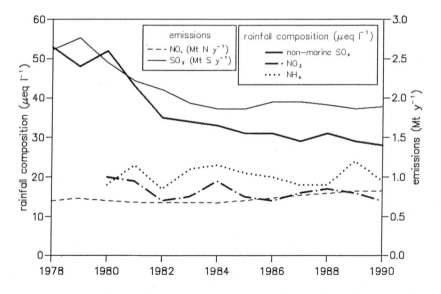

Figure 7   Trends in non-marine sulphate, nitrate and ammonium concentrations
            at Eskdalemuir, and United Kingdom SO_2 and NO_x emissions since
            1978.

and Strathclyde region where the change in S/N emission ratio may have been greater than the national average due to decreased sulphur emissions from local industry.

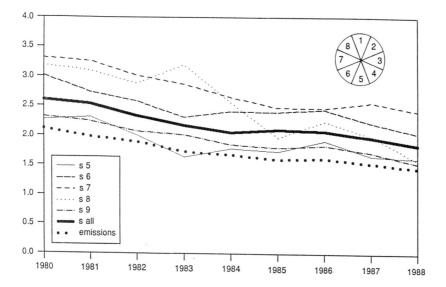

Figure 8   Trends in sulphate to nitrate ratio in precipitation composition at Eskdalemuir, stratified by sector, and in the ratio of United Kingdom SO_2 to NO_x emissions.

5 CONCLUSIONS

Current precipitation sampling techniques are subject to a number of deficiencies as discussed earlier. These include undercatch, particularly at exposed windy sites, bias of the collected sample against smaller more polluted raindrops and uncertainty as to the input of dry deposition to bulk samples. In the case of ammonium there is the additional question of losses from samples due to biological activity. As practical constraints dictate that sites are not usually located at high altitudes, corrections for seeder-feeder enhancement have to be applied. As yet these are based on a comparatively small number of investigations and extrapolation to a national scale is inevitably somewhat uncertain. Despite these limitations, data from networks of collectors have shown that wet deposition is an important pathway for the transfer of both oxidised and reduced nitrogen species to terrestrial ecosystems. Moreover data from one site in southern Scotland suggest that whereas non-marine sulphate concentrations have decreased in recent years, nitrate and ammonium concentrations have remained approximately constant.

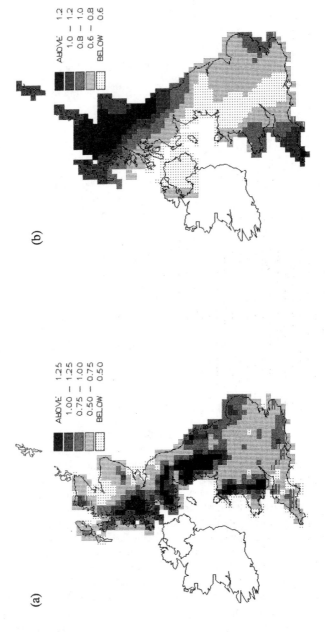

Figure 9   (a) Total wet deposited nitrogen in 1990 (gNm^{-2}) and (b) the ratio of oxidised to reduced inputs.

As illustrated in Figure 9a, current total seeder-feeder enhanced deposition ranges from less than 0.3 to greater than 2.0 gNm^{-2} over the country equivalent to 3 to 20 $kgNha^{-1}$. While the concentration patterns of ammonium and nitrate (and other major ions, notably sulphate and acidity) are broadly similar there are some differences between the two. Figure 9b plots the ratio of nitrate to ammonium over the country in 1990. In most areas ammonium deposition dominates, in places by almost a factor of two. Only in the extreme north and south-west is nitrate wet deposition greater, reflecting the respective scales of transport of ammonia and oxides of nitrogen. This is also apparent in the respective budgets as shown in Figure 10. In this the ammonia emission value must be regarded as uncertain but the contrast between the oxidised and reduced species is clear; while nitrate deposition is equivalent only to about one ninth of UK emissions, about one third of ammonia emissions are deposited within the country.

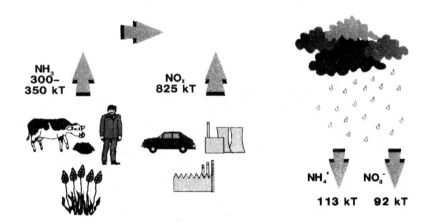

Figure 10   A partial budget of United Kingdom NH_3 and NO_x emissions and wet deposition of ammonium and nitrate.

ACKNOWLEDGEMENTS

Many of the findings reported here are based on results from the national monitoring programme funded by the Department of the Environment (contract no PECD/7/10/20). Work on precipitation collection has been funded by the European Commission, the Scottish Development Department and the Department of Trade and Industry. This support, plus the invaluable assistance of the site operators, is most gratefully acknowledged.

REFERENCES

1   B. E. Goodison, B. Sevruk and S. Klemm (1989). WMO solid precipitation measurement intercomparison: objectives, methodology, analysis. *Atmospheric Deposition*, IAHS Publ No 179, pp 57-64.

2   B. Sevruk (1989). Reliability of precipitation measurement. In *Precipitation Measurement*, Proc WMO/IAHS/ETH workshop, ETH, Zurich, B Sevruk ed. pp 13-19.

3   J. C. Rodda and S. W. Smith (1986). The significance of the systematic error in rainfall measurement for assessing wet deposition. *Atmos. Environ., 20,* 1059-1064.

4   D. J. Hall, S. L. Upton, G. W. Campbell, R. A. Waters and J. G. Irwin (1989). Further development of a snow collector for use in acid precipitation studies. Warren Spring Laboratory, Stevenage LR 752(PA).

5   D. J. Hall (1986). The precipitation collector for use in the national secondary national acid deposition network. Warren Spring Laboratory, Stevenage, LR 561 (AP).

6   B. H. Stone (1991). An assessment of the collection efficiency of United Kingdom precipitation collectors. Warren Spring Laboratory, Stevenage, LR 800 (AP).

7   S. J. Adams, S. G. Bradley, C. D. Stow and S. J. de Mora (1986). Measurements of pH versus drop size in natural rain. *Nature, 321,* 842-844.

8   J. R. Stedman, C. J. Heyes and J. G. Irwin (1990). A comparison of bulk and wet-only precipitation collectors at rural sites in the United Kingdom. *Water. Air. Soil. Pollut., 52,* 377-395.

9   Review Group on Acid Rain (1983). Acid Deposition in the United Kingdom, Warren Spring Laboratory, Stevenage.

10   R. Webster, G. W. Campbell and J. G. Irwin (1991). Spatial analysis and mapping the annual mean concentrations of acidity and major ions in precipitation over the United Kingdom in 1986. *Environ. Monit. Assess., 16,* 1-17.

11   Review Group on Acid Rain (1990), Acid Deposition in the United Kingdom, 1986-1988, Department of the Environment and Department of Transport Publications Sales Unit, South Ruislip.

12   G. W. Campbell, J. R. Stedman and J. G. Irwin (1991). Acid deposition in the United Kingdom 1989, Warren Spring Laboratory, Stevenage, LR 865(AP).

13   D. Fowler, J.N. Cape, I. D. Leith, T.W. Choularton, M. J. Gay and A. Jones (1988). The influence of altitude on rainfall composition at Great Dun Fell. *Atmos. Environ., 22,* 1355-1362.

14   A. J. Dore, T. W. Choularton, D. Fowler and A. Crossley (1992). Orographic enhancement of snowfall. *Environ. Poll., 75,* 175-179.

15 A. J. Dore, T. W. Choularton, D. Fowler and R. Storeton-West (1990). Field measurements of wet depositions in an extended region of complex topography. *Q. J. Royal Met. Soc., 116*, 1193-1212.

16 D. S. Lee, J. W. S. Longhurst, D. R. Gee and S. E. Hare (1989). Urban Acid Deposition, Results from the GMADS network, Acid Rain Information Centre, Manchester.

17 D. R. Lambert (1989). PhD Thesis, University of Leeds.

18 C. A. Johnson (1990). Report on the first years operation of the urban pollution monitoring site at London Research Station. British Gas, Research and Technology Division.

19 D. P. H. Laxen (1990). Acidity of rainfall in London. *London Environmental Supplement, 20,* 1990.

20 P. Brimblecombe and D. H. Stedman (1982). Historical evidence for a dramatic increase in the nitrate component of acid rain. *Nature, 298*, 460-462.

Modelling of Nitrogen Compounds and Their Deposition Over Europe

Hilde Sandnes, Trond Iversen, and David Simpson

MSC-W OF EMEP, THE NORWEGIAN METEOROLOGICAL INSTITUTE, PB 43 BLINDERN, 0313 OSLO 3, NORWAY

The ECE Executive Body for the convention on long-range transboundary air pollution has included the estimation of transboundary nitrogen oxides into the European Monitoring and Evaluation Programme (EMEP). The Meteorological Synthesising Centre - West (MSC-W) of the EMEP project, based in Oslo, Norway, has therefore developed a model to calculate the daily airborne concentrations and depositions of nitrogen species over Europe. The model has been used to simulate acid deposition over a period of more than 5 years (Iversen et al., 1991), and the results are being widely used in both scientific and policy discussions concerning oxidised and reduced nitrogen emissions control.

This paper will briefly describe the EMEP NO_x model, concentrating on parameterisation of the deposition processes in the model. Calculated concentrations and depositions will be illustrated. Finally, some discussion will be given of the uncertainties inherent in these modelling results, and of nitrogen deposition modelling in general.

1. The Model

The EMEP/MSC-W NO_x model is a development of previous MSC-W sulphur and nitrogen modelling activities (Eliassen and Saltbones, 1983, Hov et al., 1988, Iversen, 1990). The current model has been described in full in Iversen et al. (1991), so only a brief outline is given here. The model is Lagrangian and receptor oriented. Four times a day, trajectories for the mixed layer are calculated to arrive a set of grid points regularly spaced in a 150 km by 150 km grid covering the whole of Europe and neighbouring sea areas, as well as a set of measurement sites. Each trajectory covers a time span of 4 days and the chemistry is calculated with a time step of 15 minutes.

The model calculates concentrations, depositions and country allocated budgets for 10 chemical species : nitric oxide (NO), nitrogen dioxide (NO_2), peroxyacetylnitrate (PAN), nitric acid (HNO_3), ammonia(NH_3), ammonium nitrate, non-ammonium nitrate (hereafter NO_3^-), ammonium sulphate, sulphur dioxide (SO_2) and particulate sulphate (SO_4^{2-}). Concentrations of other species required by the model chemistry (ozone, OH and CH_3COO_2) are not calculated explicitly, but prescribed background values are specified as inputs to the model. These background values vary in general with the month of the year and the geographical position. The chemical scheme is illustrated schematically in Figure 1.

Emission data are based on the reported annual amounts from the countries participating in EMEP. There are two sources for the meteorological input data: observations and output from a Numerical Weather Prediction Model. The NWP model is developed and run operationally by the Norwegian Meteorological Institute, and produces output every 6 hours for use in the EMEP/MSC-W models.

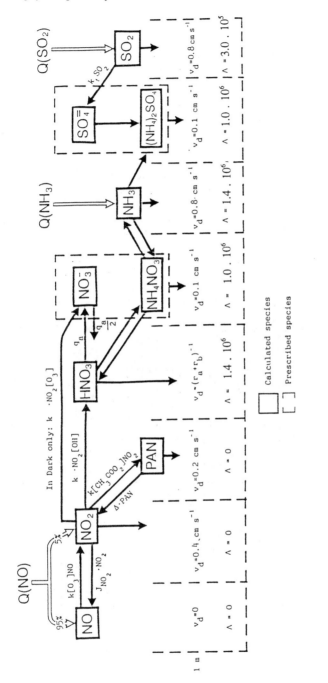

Figure 1. Overview of the chemical scheme. The 10 components inside solid boxes are calculated by the model. Species denoted in brackets are prescribed. The two dashed lines denote nitrate and sulphate particles, which both occur in two different forms in the model.

The model domain covers the whole of Europe and the output data are daily values for time periods of years. The most important determinants for the quality of the output data when designing models for this type of an air pollution dispersion model are :

1) the input data (emissions, background and meteorological data)

2) the accuracy of the transport processes (the problem of advection and convection)

3) chemistry

4) parameterisation of the dry and wet removal processes.

The model is designed in a way such that none of these processes are described in a more sophisticated way than the others, thus there is internal consistency between the different components. Weak points in our model today are problems of getting the correct gridded emission data for all components, problems connected with the vertical transport (convection) and the wet removal process.

Some recent extensions of the model.

During the last two years the model has been subjected to some important upgrading. The main changes are :

* Extension of the area where the trajectories arrive, from 720 points to 1170 points (grid and measurement points). The deposition is now calculated for an extended area including large sea areas.

* Extension of the number of emission contribution areas from 29 to 35 areas. It is now possible to allocate emissions from sub-divisions of the USSR, international ship traffic and biogenic activity over sea in the budgets.

* New treatment of precipitation (sub-grid processes). It is no longer assumed that the precipitation fill the whole grid square in a time period of 6 hours. (See section 2).

* New updated background values based partly upon measurements and partly based on results from other models. The model needs background concentrations for all of the 10 components that are integrated in the model, both in the mixed layer and the free troposphere, as well as for ozone, OH and CH_3COO_2 in the mixed layer.

* Assimilation is now in full use; this means that pre-calculated values are used as initial conditions where possible.

2. Deposition processes.

Dry deposition

Each gaseous species is assigned a basic 1-metre deposition velocity, representing the maximum value for that species (see Figure 1). These values are then modified for different latitudes and times of year, except for HNO_3 which is assumed to have the same 1-metre value everywhere. Finally, these 1-metre deposition velocities are then adjusted to values appropriate to a 50 m height using Monin Obukhov theory and surface stability parameters as given in the meteorological input. This adjustment gives a deposition flux more appropriate to the bulk of the boundary layer than that calculated using 1-metre values directly.

At night the deposition velocities are reduced by a factor of four to simulate the effects of stomatal closing in vegetation which limits the uptake of gases. Over sea areas the deposition velocities of NO_2, O_3 and PAN are set to zero.

Wet deposition

The input fields to the EMEP model include the 6-hourly precipitation intensity, P, which gives the spatially averaged precipitation for the 150 x 150 km grid square. However, in reality the precipitation varies significantly inside a grid square both in time and space. Large parts of the grid square may be completely dry. Other parts of the grid square are wet, with a precipitation intensity much larger than the grid average intensity P.

In the new model version we use results from an investigation made by a student at the University of Oslo, Mr. Per Egil Haga. He has estimated a statistical function ϕ, that gives the probability of precipitation in an arbitrary point inside a grid square as a function of the grid averaged precipitation P (see table 1). We assume that the grid square consists of a wet and a dry part in each 15 minute time-step. The probability ϕ can also be explained as the wet fraction of the grid square. If A is the total area of the grid square then the area of the wet part is $\phi*A$ and the area of the dry part is $(1-\phi)*A$. The precipitation intensity in the wet part is $P_w = P/\phi$ so that the total volume of precipitation is the same if we use the grid averaged precipitation intensity over the area A, or if we use the sub-grid precipitation intensity P_w over the area $\phi*A$. The pollutant concentration inside a grid with precipitation is now treated as a linear function of one part of the concentration $\phi*Q$ that is hit by precipitation and another part $(1-\phi)*Q$ that is in the dry part of the grid area.

Table 1. The fraction ϕ, of a grid square area which is hit by precipitation as a function of the six hourly grid square averaged precipitation amount P.

P (mm/6h)	0	3	6	9	12	20	50	90	150
ϕ (%):	0	31	48	60	66	72	80	85	91

In the old model version where we used the grid averaged precipitation P directly, the air parcel encountered precipitation more frequently, but with a smaller intensity than in reality. The consequences of the parameterisation of the subgrid precipitation described above are that the frequency of being hit by precipitation and the precipitation intensity are closer to reality. A component with high wet removable rate will now have a larger probability of being transported to longer distances.

3. Results and comparison with observations

Details of the EMEP measurement programme can be found in several routine reports from EMEP/CCC (eg. Schaug et al., 1991). Unfortunately, only a few of the components calculated are measured regularly with a reasonable coverage. In this paper we will only consider NO_2 in air and both nitrate and ammonium in precipitation. Results for other species and years are given in Iversen et al. (1991).

Note that in presenting comparisons in air a 75% data coverage is required from a measurement station for it to be included in the comparison. For species in precipitation, the requirement is that measured concentrations exist on at least 25% of the days in which the model has precipitation. This apparently low figure arises from the fact that the model's grid average precipitation data overestimates the frequency of precipitation events as compared to a single measurement station, as discussed in section 2.

NO_2

Figure 2 illustrates the comparison of the modelled versus observed annual NO_2 concentrations at all sites in 1989. The model underpredicts measured concentrations by a factor of 2 on average. Although the underpredictions at some sites can be attributed to unrepresentative measuring stations or incorrect emissions (Iversen et al., 1991), there are undoubtedly some systematic reasons for this underprediction which are difficult to avoid. These include the measurement of pollution from local ground-level sources of NO_2 (eg. from soils or vehicles, Simpson, 1991), and substantial uncertainties in measurements at low concentrations (Fehsenfeld et al., 1988). For these reasons, the comparison of modelled versus observed NO_2 concentrations is of only limited value in evaluating model performance. Better observations, and more knowledge about the contributions of natural and other local sources to ground level concentrations are required before much further progress can be made in this evaluation.

$NO_3^-{}_{(l)}$

The corresponding results for nitrate in precipitation (Figure 3) show a large scatter, but the mean levels and correlation for this component are quite good. There is a tendency to underestimate at low concentrations, probably associated with the lack of a background (tropospheric) contribution to wet depositions (Iversen et al., 1991). The results for two UK sites are presented in Figure 4. In general, the model seems to do rather well for this species.

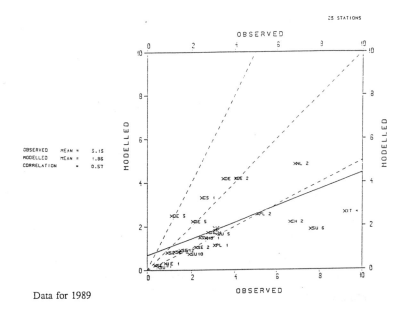

Data for 1989

Figure 2. Scatter plots of modelled and observed concentrations of NO$_2$ for a selection of stations with a data coverage of more than 75%. Units : μg(N) m^{-3}. The dashed lines show perfect agreement and disagreement with a factor of two. The full line represents optimal linear regression, i.e. the sum of squares of distances measured from the points normal to the line is minimised.

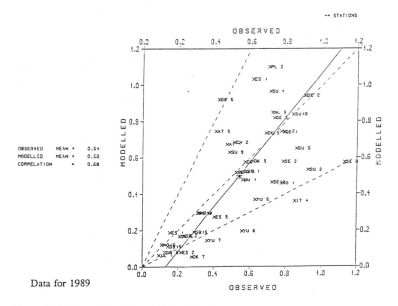

Data for 1989

Figure 3. Scatter plots of modelled and observed concentrations of nitrate in precipitation for a selection of stations with a data coverage of more than 25%. Units : mg(N) l^{-1}.(For explanation of choice of data coverage see text).

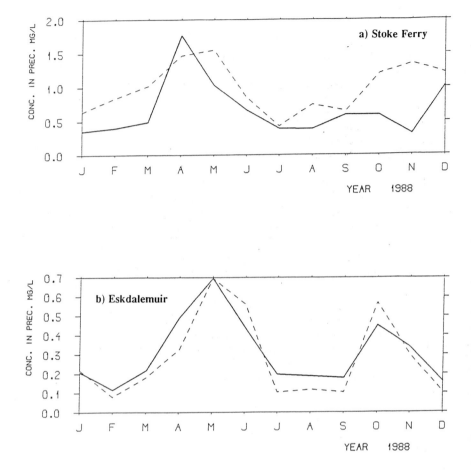

Figure 4. Measured versus observed monthly concentrations of nitrate in precipitation at two UK sites for 1988. Units: mg(N) l⁻¹. The solid line represents observed values, and the dashed line modelled values.

NH_4^+ (l)

The results for ammonium in precipitation (Figure 5) are seen to be similarly good. Given the large uncertainties in the emissions of NH_3, and the unquantifiable contribution of local emission sources to measurements, these results are, if anything, better than might be expected.

Overall, given the large uncertainties in the emissions and measurement quality, the model results are satisfactory. Many of the discrepancies between modelled and observed concentrations can be explained in terms of known shortcomings in either the model (eg. the large grid-size, lack of vertical resolution) or in the input data.

As an example of a common application of EMEP models, Figure 6 shows the deposition of oxidised nitrogen in 14 countries which can be attributed to emissions from the United Kingdom (UK). These results are calculated firstly with the original ("old") model version, and secondly with the updates described above included ("new" model). Not surprisingly, the UK is the largest receiver of nitrogen oxides emitted in the UK. However, this diagram also shows substantial contributions to France (FR), Germany (DE), and Norway (NO).

Comparing the new with the old model version, we see that the results are fairly similar in both cases. The deposition from UK to itself has decreased somewhat. The main trend for the deposition from the UK to other countries is an increased deposition. This means that pollution are transported longer distances before it is deposited in the new model version. This is partly due to the new treatment of precipitation, and partly due to changes for the coefficients for wet and dry removal for some of the chemical components.

In Iversen et al.(1991) the contribution of each country's emissions to the deposition fields for every other country is presented for a period of 5 years. For countries such as Norway, it is found that long-range transport makes up the major part of the calculated nitrogen deposition within the country. These, and previous EMEP calculations, have clearly demonstrated that nitrogen compounds, especially nitrogen oxides, have transport scales comparable to those of sulphur, and that such European scale modelling is essential to understand the fate of emitted nitrogen species.

4. Uncertainties in model output

In considering the results of any air pollution model, it is important to understand the sources and possible magnitude of any uncertainties in the results, especially if these results are to be used in emission control evaluations. As well as giving information on the current accuracy of a model's predictions, this analysis should indicate where future improvements can be made in the model formulation or input data.

Sensitivity to parameter changes

A number of sensitivity tests have been carried out using an earlier version of the EMEP NO_x model, to evaluate the effects of various assumptions on the predicted concentrations of

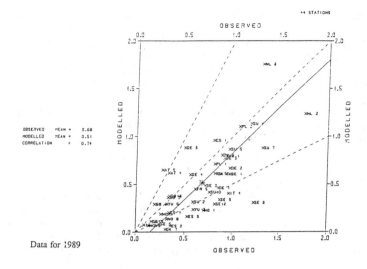

Data for 1989

Figure 5. Scatter plots of modelled and observed concentrations of ammonium in precipitation for a selection of stations with a data coverage of more than 25%. Units : mg(N) l⁻¹. (For explanation of choice of data coverage see text).

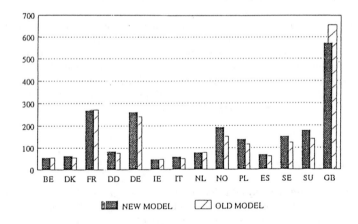

Figure 6. Contribution of United Kingdom emissions to oxidised nitrogen deposition in other countries, 1988. Unit : 100 tonnes as N per annum.

nitrogen species at 7 measurement sites in the UK (Simpson, 1990). Although these results strictly apply to a model similar to the "old" version described above, these results do illustrate the response of a long-range transport model to some of the most important model parameters. Table 2 presents a summary of these results.

Although Table 2 indicates that most of these sensitivity tests can have substantial effects on particular NO_y components, only two parameters were found to produce a dramatic change in the total oxidised nitrogen deposition at the UK sites: NO_x emissions and the wet deposition parameterisation.

Increasing or decreasing NO_x emissions by 50% is seen to produce an almost linear response in the predicted concentrations and depositions of oxidised nitrogen species. The only significant non-linearity occurs with the wet deposition, as a result of the significant contribution of background NO_y species to this component. The uncertainty of the model output is therefore directly related to the uncertainties in the quality of the NO_x emissions input data.

Increasing the ammonia emission by 100% has a dramatic effect on the HNO_3 concentrations, but has little effect on the other concentrations. As HNO_3 makes only a small contribution to the NO_y budget, the calculated depositions are only slightly affected by even such a large change in the NH_3 emissions. The most important effect of uncertainties in NH_3 emissions then will be in the calculation of reduced nitrogen deposition.

It is interesting that although changing the scavenging ratios can have a moderate effect on wet deposition, but the total deposition is hardly affected. The reason for this lack of sensitivity is that the soluble nitrogen species are very rapidly removed by rain with almost any reasonable value of the scavenging co-efficient - it makes very little difference to the model results if the removal takes place over a period of one or six hours.

The last sensitivity test (wet deposition only in last 6 h) requires some explanation. In the model version used in these tests, uniform rainfall was assumed over a grid square (in contrast to the probabilistic treatment described in section 2), so that if rain was measured at any of the stations in a grid square the model would assume rainfall throughout the square. This assumption led to a too-frequent occurrence of rainfall and consequently reduced the lifetime of NO_y in the boundary layer. In this last sensitivity test, the extreme assumption was made that rainfall had no effect on concentrations except during the last six hours, ie. until the trajectory was over the receptor point. These two extreme assumptions (P constant over grid square, or P=0 until the last 6 hours) are seen to give markedly different wet depositions (> 200%), and to have a great influence on the calculated total depositions. Although the current model version uses a more realistic description of the wet-deposition process, this area is obviously a source of great uncertainty. Unfortunately, the processes involved in wet scavenging and deposition are so complex and little understood that it will be some time before any model, no matter how complex, can produce significantly more reliable results.

Table 2 Summary of sensitivity analysis for early NO_x model results: percentage changes in calculated concentrations and depositions of oxidised nitrogen. Average over 7 UK sites.

Sensitivity test:		NO_2	NO_3	HNO_3	PAN	Wet-N Dep.	Dry-N Dep.	Total-N Dep.
NO_x Emissions	+ 50%	49	45	52	47	36	48	42
	- 50%	-48	-46	-42	-46	-36	-46	-41
NH_3 Emissions	+100%	0	14	-63	0	1	-12	-5
UK emission grid:	50 km	2	11	-11	6	8	2	5
UK emission grid:	30 km	0	10	-9	6	8	1	5
NO_2 dep. velocity	÷ 2	5	5	5	6	3	-16	-6
HNO_3 dep. velocity	x 2	0	-3	-22	0	-6	11	2
	÷ 2	0	3	21	0	6	-8	0
Initial Concs.	x 2	1	6	14	6	27	7	17
Initial Concs.	÷ 2	0	-2	-6	-2	-12	-3	-8
Background concs.	÷ 2	31	-27	-28	-36	-19	-3	-11
Mixing height	+ 25%	-17	-13	3	-17	6	-9	-1
Mixing height	- 25%	26	22	-1	28	-3	17	6
vertical velocity	x 2	7	-3	-14	3	-1	-1	-1
vertical velocity	÷ 2	-2	2	11	0	3	2	3
No $NO_{3.}^{(p)} \rightarrow HNO_3$ pathway		0	12	-58	0	10	-11	0
Scavenging ratio	x 2	0	-5	-7	0	14	-3	5
	÷ 2	0	6	8	0	-15	4	-5
Wet deposition only in last 6 h		0	24	32	0	245	19	133

NB. all of these experiments were conducted with an early version of the NO_x model, so the numbers given are only intended as an illustration of the behaviour of this type of model. In particular, the wet deposition parameterisation is different in the new model to that used here. The wet-N, dry-N, and total-N columns refer to depositions of oxidised nitrogen. For full explanation, see Simpson (1989).

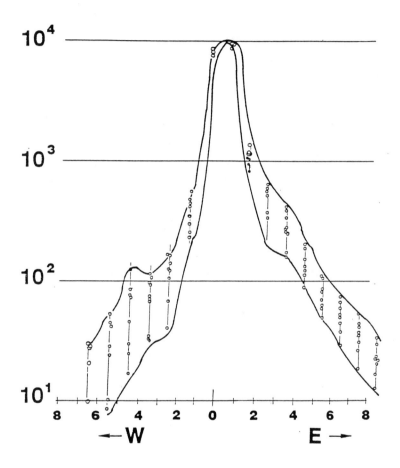

INTERANNUAL VARIABILITY OF S-DEPOSITION

Figure 7. A cross-section through grid squares in the E-W direction of calculated annual deposition patterns arising from emissions in the former German Democratic Republic. The deposition values for each of the grid squares is given for eight consecutive years (1979-1987) by the small circles. The two curves denote the upper and lower bounds of the annual depositions. Units : mg(S) m^{-2} , note that the scale is logarithmic. Since E-W is diagonal in the grid, the distance unit is approximately 150 km x $\sqrt{2}$ = 212 km.

Inter-annual variability

The level and patterns of deposition arising from a particular emission area depend very much on the meteorological conditions, even when we are averaging over periods as long as one year. To illustrate this, an experiment has been conducted with the EMEP sulphur model (Eliassen et al., 1990), in which the depositions from the former German Democratic Republic were calculated for eight consecutive years. This country was chosen because it only covers a few grid squares. The annual sulphur depositions calculated along an East - West transect passing through this country are illustrated in Figure 7.

This diagram illustrates that the annual deposition arising from a small area to a certain EMEP grid square can vary by a factor from 2 to 10 in the course of eight years, depending on the relative position of the grid-square. The smaller the size of the emission area considered in a study such as this, the larger such inter-annual variability will be. This experiment strongly reinforces the need for models which deal with meteorology in a realistic way, and places a limit on the smallest size of emission area which it is sensible to consider in evaluating policy options.

7. Conclusions

This paper has briefly described the EMEP MSC-W NO_x model, which has been developed in order to describe the long-range transport of both oxidised and reduced nitrogen compounds in Europe. The model attempts to describe the most important physical and chemical processes, namely emissions, transport, chemistry and removal, in a balanced and realistic way. Comparison with observations suggest that the model reproduces these processes in a satisfactory way, although limitations in both the model and the quality and coverage of the measurement data prevent a through assessment at this stage.

Sensitivity analysis has shown that the main uncertainties in the model output are likely to arise from the specification of the emissions data, and the parameterisation of the wet deposition process. Both of these problems are likely to be common to any air pollution model for nitrogen species.

It was also shown that because of meteorological variability, the depositions from a relatively small area (~300 x 300 km^2), even when averaged over a full year, can vary by almost an order of magnitude from one year to the next at particular receptors. This variability must be taken into account if we attempt to assess optimised control policies for nitrogen species in Europe.

9. Acknowledgements

The EMEP MSC-W model has been developed over a number of years with contributions from several sources. In addition to the authors, this includes Anton Eliassen, Nina Halvorsen, Sophia Mylona and Jørgen Saltbones of MSC-W, and Øystein Hov from the University of Bergen, Norway. The preparation of meteorological data from the numerical weather prediction model was by Thor Erik Nordeng of the Norwegian Meteorological Institute. The development of the EMEP NO_x model has received funding through EMEP, the Nordic Council of Ministers, the UK Dept. of the Environment, and the Government of Canada.

6. References

Eliassen, A. and Saltbones, J. (1983) Modelling of long-range transport of sulphur over Europe: A two year model run and some model experiments. Atmos. Environ., 17, 1457-1473.

Eliassen, A., Saltbones, J., and Sandnes, H. (1990) Some properties of deposition patters of nitrogen and sulphur and their implications for European abatement strategies. EMEP MSC-W Note 1/90.

Fehsenfeld, F.C., Parrish, D.D., and Fahey, D.W. (1988) The measurement of NO_x in the non-urban troposphere. In Tropospheric ozone, ed. I.S.A.Isaksen, Reidel Publishing Company, 1988.

Hov, Ø., Eliassen, A. and Simpson, D. (1988) Calculation of the distribution of NO_x compounds in Europe. In Tropospheric ozone, ed. I.S.A.Isaksen, Reidel Publishing Company, 1988.

Iversen, T. (1990) Calculations of long-range transported sulphur and nitrogen over Europe. in The Science of the Total Environment, 96, 87-99, Elsevier, 1990.

Iversen, T., Halvorsen, N.E., Mylona, S. and Sandnes, H. (1991) Calculated budgets for airborne acidifying components in Europe, 1985, 1987, 1988, 1989 and 1990. EMEP/MSC-W Report 1/91.

Schaug, J., Pedersen, U., and Skjelmoen, J.E. (1991) Data report 1989. Part 1: Annual summaries. EMEP/CCC Report 2/91. The Norwegian Institute for Air Research, Lillestrøm, Norway.

Simpson, D. (1989) Application of the EMEP NO_x model to the United Kingdom. Stevenage, Warren Spring Laboratory Report LR 719 (AP).

Simpson, D., Perrin, D.A., Varey, J.E., and Williams, M.L. (1990) Dispersion modelling of nitrogen oxides in the United Kingdom., Atmos. Environ., 24A, No. 7, 1713-1734.